Chemical Calculations

Paul Yates

Lecturer in Physical Chemistry
Department of Chemistry
Keele University
Keele
UK

BLACKIE ACADEMIC & PROFESSIONAL

An Imprint of Chapman & Hall

London · Weinheim · New York · Tokyo · Melbourne · Madras

Published by
Chapman & Hall, 2–6 Boundary Row, London SE1 8HN, UK

Chapman & Hall, 2–6 Boundary Row, London SE1 8HN, UK

Chapman & Hall, GmbH, Pappelallee 3, 69469 Weinheim, Germany

Chapman & Hall USA, 115 Fifth Avenue, New York, NY 10003, USA

Chapman & Hall Japan, ITP-Japan, Kyowa Building, 3F, 2-2-1 Hirakawacho, Chiyoda-ku, Tokyo 102, Japan

Chapman & Hall Australia, 102 Dodds Street, South Melbourne, Victoria 3205, Australia

Chapman & Hall India, R. Seshadri, 32 Second Main Road, CIT East, Madras 600 035, India

First edition 1997

© 1997 Chapman & Hall

Typeset in 10/12 Times by Blackpool Typesetting Services Limited, UK
Printed in Croatia by Zrinsky d.d., Cakovec

ISBN   0 412 74890 8

A catalogue record for this book is available from the British Library

Library of Congress Catalog Card Number: 96-80205

Chemical Calculations

For Julie, Catherine and Christopher

Contents

Preface

Students arriving at university to study chemistry are frequently surprised, not to say disappointed, when they realize that it will be necessary for them to study mathematics alongside their chosen subject, at least during the first year. The need to do this will be justified by staff who will say that a reasonable knowledge of mathematics is essential for a study of physical chemistry. This is quite true, but I think that if we look a little more closely there is a more fundamental reason underpinning such statements. It is that chemistry is a physical science, obeying laws and principles which can only be described within the framework of mathematics. This is as true of organic and inorganic as it is of physical chemistry, although certainly not so obviously.

If you look at the contents page of this book, you will see that the chapters are broken down into much the same topics as you will find in a textbook on physical chemistry. This is not a chemistry textbook however, and no attempt is made to teach any of the physical chemistry which is available in many other textbooks. My aim in writing this book is to explain the mathematical techniques which accompany each of these chemical topics, and to show that, even when the mathematics starts to become involved, the chemistry is not far away. Because of this emphasis, my aim has been to include more detail than is possible in physical chemistry textbooks, where to do so would be to lose sight of the overall chemistry. The organization is therefore different to that of many mathematics books aimed at chemists, and a consequence of this is that a number of mathematical ideas are met on more than one occasion.

I have tried to include as many examples as possible. Within the text these are called Worked Examples and are accompanied by detailed solutions. At the end of each chapter you will find short maths-based Exercises to work through, as well as longer, chemically related Problems. Answers to these are included at the back of the book, and I have tried to make these sufficiently detailed to be of help to a student working independently. There is also an appendix on Fundamentals which contains information on some of the absolute basics of mathematics, without which the more advanced aspects cannot be appreciated.

The material contained in this book will, I hope, lend itself to being used to accompany the mathematics courses which are now almost universally provided for chemists at degree level. However, I also hope that the structure of the book will make it useful as a companion throughout a study of physical chemistry, so that the mathematics of a particular topic will be to hand when required.

One thing I have not attempted to do is to address some of the very exciting current developments in mathematical chemistry. As I write this, the current issue of *Chemistry in Britain* contains an article entitled 'Mathematics with

Molecules' by Dennis Rouvray, which speculates that mathematicians will be turning to chemists for help with their problems before long. The need for chemists to have a reasonable knowledge of mathematics is clearly very strong.

Acknowledgements

The author of any book receives help in various ways, some more obvious than others. Dominic Recaldin, formerly the Director of the Textbook Unit at Chapman & Hall, first encouraged me to pursue this project and was instrumental in helping me to formulate the plan of this book; my thanks go to him and also to Jon Walmsley for his assistance in completing my manuscript. Most of the material contained here was tested on chemistry students at Keele University over the last three years or so, and their feedback has been immensely valuable. Finally, I would like to thank my wife Julie, who cast a critical non-scientist's eye over the manuscript and who has given me unending support and encouragement over the last few years.

Paul Yates
Keele
June, 1996

Experimental techniques 1

Despite the increasing use of computers in many aspects of chemistry, and the enormous advances in theoretical areas such as quantum mechanics, most chemists still perform laboratory experiments. Particularly in physical chemistry, the object of an experiment is often to collect some form of numerical data which is then analysed to determine the value of some physical quantity. It is important that you are able to handle this sort of data. At first sight this may seem trivial, but in some ways the appropriate treatment of data is one of the most difficult areas of mathematical chemistry to master. Every case seems to be different, and it is only by gaining experience of several calculations that you will become confident in performing this data analysis.

1.1 Measurement in chemistry

Many of the measurements made in physical chemistry are of quantities which are easy to visualize. For example, a burette could be used to measure a volume, and a thermometer could be used to measure temperature. Other measurements which you will make in an undergraduate chemistry course will be of less familiar quantities. Absorbance is measured using a spectrophotometer and electromotive force is measured using a digital voltmeter, for example. Frequently, the values read from these instruments will be used in subsequent calculations, which will usually be performed with the help of a calculator. Calculators do allow you to perform complex calculations easily, but unfortunately they do not always give results which can be quoted directly to give a realistic answer.

Let us take a simple example to illustrate this. Suppose you wish to standardize a solution of sodium hydroxide which has a concentration of approximately $0.1 \ mol \ dm^{-3}$. To do this, you titrate $25 \ cm^3$ of the sodium hydroxide with hydrochloric acid which you are told has a concentration of exactly $0.0994 \ mol \ dm^{-3}$. Suppose that $25 \ cm^3$ of the sodium hydroxide is neutralized by $24.85 \ cm^3$ of hydrochloric acid. The concentration of sodium hydroxide, $[NaOH]$, is then given by the formula

$$[NaOH] = \frac{24.85 \ cm^3 \times 0.0994 \ mol \ dm^{-3}}{25 \ cm^3}$$

Using the calculator, this gives a value for $[NaOH]$ of $0.0988036 \ mol \ dm^{-3}$. The question we need to ask ourselves is 'are we justified in quoting the final answer to this many figures?'. You may guess that, in this case, the answer is 'no', but

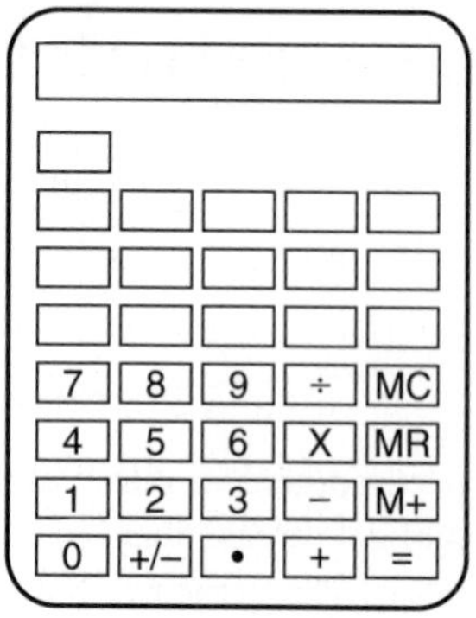

Figure 1.1   The calculator which will be used to show how to perform certain calculations.

why is that and how can we decide what number of figures would be reasonable? To do this we need to understand the concept of **significant figures**.

Incidentally, in this book, you will be shown the buttons you need to press on your calculator to perform some of the calculations which are less obvious. When you see the calculator icon ▦ in the text, the corresponding series of calculator buttons to press will be shown in the margin, using the picture of a calculator pad shown in Figure 1.1.

1.1.1   Significant figures

It is easy to confuse the concepts of decimal places and significant figures. In the example above, we could quote the final value for [NaOH] to any number up to 7 decimal places. For example,

$$[\text{NaOH}] = 0.0988 \text{ mol dm}^{-3} \text{ to 4 decimal places}$$

$$[\text{NaOH}] = 0.098\,80 \text{ mol dm}^{-3} \text{ to 5 decimal places}$$

$$[\text{NaOH}] = 0.098\,8036 \text{ mol dm}^{-3} \text{ to 7 decimal places}$$

The number of significant figures in each expression is, however, quite different from the number of decimal places. To count these, we need to apply the following rules:

1. Ignore any leading zeros (such as those in 0.001, for example).
2. Include any zeros 'inside' the number (such as the second zero in 0.101, for example).
3. Use your judgement to decide whether trailing zeros are significant or not (they would be, in a burette reading of 25.00 cm^3, for example, since this is clearly different from a reading of, say, 25.05 cm^3).

So, we now have for the above expressions of sodium hydroxide concentration,

$$[\text{NaOH}] = 0.0988 \text{ mol dm}^{-3} \text{ to 3 significant figures}$$

$$[\text{NaOH}] = 0.098\,80 \text{ mol dm}^{-3} \text{ to 4 significant figures}$$

$$[\text{NaOH}] = 0.098\,8036 \text{ mol dm}^{-3} \text{ to 6 significant figures}$$

Now consider what happens if we want to give the concentration to 5 significant figures. We might expect a value of $0.098\,803 \text{ mol dm}^{-3}$, but if we look at the figure immediately to the right of this which has been deleted, we find that it is a 6. This means that our value should be closer to $0.098\,804 \text{ mol dm}^{-3}$, and so we need to 'round up' the last digit of our expression.

When quoting a value to a specified number of decimal places or significant figures, always look at the first digit you are planning to delete. If it is between 5 and 9 round the previous digit up; otherwise leave the previous digit as it is.

One case in which application of the rules of significant figures can seem a little strange is when you are asked to round a number greater than 10 to fewer figures than there are digits present before the decimal point, such as 1265 to 2 significant figures. In these cases, we need to fill out the remaining places with zeros; 1265 becomes 1300 using the usual rounding rules. Remember that when going the other way, these zeros are not significant.

Worked example 1.1

Write each of the following quantities to (i) 3 significant figures and (ii) 2 decimal places:

(a) a volume of 17.927 cm^3
(b) an energy change of $396.748 \text{ kJ mol}^{-1}$
(c) a rate constant of $1.0846 \text{ dm}^3 \text{ mol}^{-1} \text{ s}^{-1}$
(d) the ideal gas constant $0.082\,0578 \text{ dm}^3 \text{ atm K}^{-1} \text{ mol}^{-1}$

Chemical background

(a) We have already seen that measurement of volume is important in chemistry in volumetric analysis. Volume changes are also important in thermodynamics where much use is made of the properties of gases.
(b) Energy changes accompany chemical reactions and can be predicted using thermodynamic methods. They can be measured experimentally using one of various types of calorimeter available, such as that shown in Figure 1.2.
(c) The rate constant determines how fast a particular chemical reaction will take place. You may have noticed that the units can be thought of as per concentration per time, and it is the concentration which often governs the rate of a chemical reaction at a given temperature. As we will see in

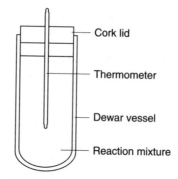

Figure 1.2   Schematic diagram of a Dewar calorimeter.

Chapter 4, the rate constant has different units depending on the type of reaction to which it refers.

(d) We usually express the ideal gas constant in units of $J\,K^{-1}\,mol^{-1}$, but the form given in this example can be useful when pressure values are given in units of atm. The ideal gas constant occurs in many branches of physical chemistry, and not just those concerned with the properties of gases!

Solution to worked example

(a) Counting the first 3 significant figures is straightforward and gives a value of 17.9 cm^3. To 2 decimal places, we obtain 17.92 cm^3, but a 7 has been deleted (between 5 and 9) so we need to round up the last digit to give 17.93 cm^3.

(b) The first 3 significant figures take us as far as the decimal point, but the first digit deleted is 7 (between 5 and 9), so we round from 6 to 7 to give 397 $kJ\,mol^{-1}$. To 2 decimal places, we have 396.74 $kJ\,mol^{-1}$, but the first digit deleted is 8 so we round up to get 396.75 $kJ\,mol^{-1}$.

(c) The first 3 significant figures give 1.08 $dm^3\,mol^{-1}\,s^{-1}$; notice that the zero inside the number is included as a significant figure. The first digit deleted is 4, so no rounding is required. The result is the same if we round to 2 decimal places.

(d) We ignore the leading zeros, and count 3 significant figures to give 0.0820 $dm^3\,atm\,K^{-1}\,mol^{-1}$. However, the first digit deleted is 5, so we round up to give 0.0821 $dm^3\,atm\,K^{-1}\,mol^{-1}$. Counting 2 decimal places gives 0.08 $dm^3\,atm\,K^{-1}\,mol^{-1}$, and as a 2 has been deleted no additional rounding is required.

1.1.2   Combining quantities

Having decided how many significant figures are appropriate for each quantity in a calculation, you can determine the appropriate number of significant figures to which to quote the final value calculated from some combination of them. You need to apply the following rules.

1. When *adding* or *subtracting* numbers, the answer should be given to the smallest number of *decimal places* in the quantities, so that we write $1.23 + 4.5 \simeq 5.7$. for example rather than $1.23 + 4.5 = 5.73$.
2. When *multiplying* or *dividing*, the answer should be given to the smallest number of *significant figures* in the quantities, so that

$$\frac{7.432}{2.1} \simeq 3.5$$

for example, rather than

$$\frac{7.432}{2.1} = 3.539$$

Worked example 1.2

Find the values of the following expressions, giving the appropriate number of figures in your answer.

(a) The difference in burette readings of 36.35 cm^3 and 11.2 cm^3.
(b) The density of a solution calculated from the formula

$$\text{density} = \frac{\text{mass}}{\text{volume}} = \frac{17.098 \text{ g}}{6.2 \text{ cm}^3}$$

(c) The concentration of a solution given by the expression

$$\frac{25.15 \text{ cm}^3 \times 0.104 \text{ mol dm}^{-3}}{25.00 \text{ cm}^3}$$

Chemical background

(a) This is actually an example of bad laboratory practice. The burette should always be read to the same degree of precision, which would normally be to $\pm 0.05 \text{ cm}^3$. This is true even if the starting value was exactly zero, so this would be given as 0.00 cm^3 rather than 0.0 cm^3. Otherwise, notice that the rule would apply exactly as in this case.
(b) The density of a substance is found by dividing its mass by its volume. One example of doing this in the laboratory, for solutions, is in an experiment to determine partial molar volumes.
(c) This is simply another example of the standardization procedure met earlier. It is important to realize that the division by 25.00 cm^3 refers to a pipette volume, so the trailing zeros are important to give the precision of the pipette.

Solution to worked example

(a) The subtraction $(36.35 - 11.2)\,cm^3$ on a calculator gives the result $25.15\,cm^3$. However, we see that the second of the two values is given to only 1 decimal place, so the final answer would be $25.1\,cm^3$. However, the digit 5 has been deleted, so we need to round up to give $25.2\,cm^3$.

(b) Using the calculator gives an answer of $2.757\,741\,9\,g\,cm^{-3}$ which clearly contains more figures than we need. If we look at the two quantities given in the question, we see that they have 5 and 2 significant figures respectively. Giving this answer to 2 significant figures gives $2.7\,g\,cm^{-3}$, but we have deleted the digit 5 and so need to round this up to $2.8\,g\,cm^{-3}$.

(c) The calculator again gives a large number of figures, and we obtain an answer of $0.104\,624\,mol\,dm^{-3}$. Looking at the three quantities in the question shows that they are given to 4, 3 and 4 significant figures respectively. (We include in this count the zero inside the second quantity, and the trailing zeros in the pipette volume.) The answer therefore needs to be given to 3 significant figures: $0.104\,mol\,dm^{-3}$. Once again, rounding up is required because a 6 has been deleted and the final answer should be quoted as $0.105\,mol\,dm^{-3}$.

1.2
Stoichiometric calculations

In the previous section, we met the idea of calculations using the results of volumetric analysis. The reaction between sodium hydroxide and hydrochloric acid takes place in a simple $1:1$ ratio, so we did not have to consider this aspect of the reaction any further. Many chemical reactions are more complicated than this, and it is useful to be able to calculate the amount or mass of products from a given amount or mass of the reactants.

1.2.1   Multiplication and division by an integer

In previous examples of chemical equations, we have been dealing with real numbers where it has been necessary to determine the number of significant figures of each quantity involved. However, when multiplying or dividing by an integer, we can ignore the number of significant figures suggested by the integer. Integers are *exact* numbers, so that 4 could be thought of as 4.0000000 ..., for instance, to ensure that it is treated as having more significant figures than any other quantity in the calculation.

Worked example 1.3

In the reaction

$$Fe_2O_{3(s)} + 3CO_{(g)} \rightarrow 2Fe_{(1)} + 3CO_{2(g)}$$

calculate the mass of iron produced when 503 g of carbon monoxide is passed over an excess of Fe_2O_3. The molecular mass of CO is $28.01\,g\,mol^{-1}$ and the atomic mass of iron is $55.85\,g\,mol^{-1}$.

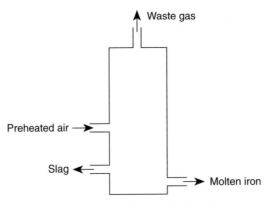

Figure 1.3   Schematic diagram of a blast furnace.

Chemical background

This equation represents the reaction which takes place in a blast furnace (Figure 1.3), where iron is extracted from iron ore by reduction with carbon monoxide. Coke is burned to form carbon dioxide which then reacts with more hot coke to form carbon monoxide.

Unlike the previous examples we have met, where equal numbers of molecules of each reactant combine, here we see that 1 molecule of iron oxide combines with 3 molecules of carbon monoxide. The fact that these numbers are still integers is a consequence of Dalton's atomic theory expressed in the law of multiple proportions which was formulated in 1803.

A discussion of the significance of Dalton's atomic theory can be found in *Chemistry in Britain*, **30**, 920–1, 1994.

Solution to worked example

It is important to realize that the coefficients which appear in this equation refer to the relative numbers of molecules which combine in the reaction. These are known as the amounts of each species and are measure in moles. Using the relationship

$$\text{amount} = \frac{\text{mass}}{\text{molecular mass}}$$

and substituting the given values into this equation, we have

$$\text{amount of CO} = \frac{503 \text{ g}}{28.01 \text{ g mol}^{-1}}$$

and

$$\text{amount of Fe} = \frac{\text{mass of Fe}}{55.85 \text{ g mol}^{-1}}$$

From the reaction equation, we see that the ratio amount of CO : amount of Fe is
3 : 2, which is equivalent to the equation

$$\frac{\text{amount of CO}}{\text{amount of Fe}} = \frac{3}{2}$$

We can rearrange this to give

$$\text{amount of Fe} = \left(\tfrac{2}{3}\right) \times \text{amount of CO}$$

Substituting in our expression for the amounts of each of these gives

$$\frac{\text{mass of Fe}}{55.85 \text{ g mol}^{-1}} = \left(\frac{2}{3}\right) \times \frac{503 \text{ g}}{28.01 \text{ g mol}^{-1}}$$

One final rearrangement allows us to calculate the mass of Fe to be

$$55.85 \text{ g mol}^{-1} \times \left(\frac{2}{3}\right) \times \frac{503 \text{ g}}{28.01 \text{ g mol}^{-1}}$$

Evaluating this on a calculator gives the value 668.631 44 g. We can ignore the
integers 2 and 3 in considering the appropriate number of significant figures, and
are left with 4, 3 and 4 figures respectively in the remaining quantities. It is
therefore appropriate to give our answer to 3 significant figures, so our final
answer should 669 g (we have rounded up the final digit as the first digit deleted
is 6 which is greater than 5).

It is worth noting that the calculator was only used at the very end of the
calculation. Had it been used in the intermediate steps, we might have ended up
converting 2/3 to 0.666 66 … which would have caused difficulties at the end
when trying to determine the correct number of significant figures.

Worked example 1.4

What volume of sulphuric acid of concentration 0.0502 mol dm^{-3} is required to
neutralize 25.00 cm^3 of 0.0995 mol dm^{-3} sodium hydroxide?

Chemical background

Sulphuric acid is known as a dibasic acid because it contains two hydrogen
atoms which may dissociate from the molecule to give the hydrogen ions which
confer acidity (Figure 1.4). In hydrochloric acid, there is only one such hydrogen and it is known as a monobasic acid.

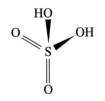

Figure 1.4 Structure of sulphuric acid, H_2SO_4.

Solution to worked example

The equation for the neutralization reaction taking place is

$$2NaOH + H_2SO_4 \rightarrow Na_2SO_4 + 2H_2O$$

so we see that only 1 mol of sulphuric acid is required for every 1 mol of sodium hydroxide. Since the ratio amount of sodium hydroxide : amount of sulphuric acid is 2 : 1 we can write

$$\frac{\text{amount of sodium hydroxide}}{\text{amount of sulphuric acid}} = \frac{2}{1}$$

and then substitute the values into this equation to give

$$\frac{25.00 \text{ cm}^3 \times 0.0995 \text{ mol dm}^{-3}}{\text{volume of sulphuric acid} \times 0.0502 \text{ mol dm}^{-3}} = \frac{2}{1}$$

This rearranges to give the equation

$$\text{volume of sulphuric acid} = \frac{25.00 \text{ cm}^3 \times 0.0995 \text{ mol dm}^{-3}}{0.0502 \text{ mol dm}^{-3} \times 2}$$

which, on using a calculator, yields a volume of 24.775 896 42 cm^3. To determine the number of significant figures in this answer, we have to realize that the volume of sodium hydroxide will have been determined by a pipette and so the trailing zeros given will be significant. Ignoring the integer value of 2, we are left with combining 25.00 cm^3 which has at least 4 significant figures, 0.0995 mol dm^{-3} which has 3 significant figures (ignore leading zeros) and 0.0502 mol dm^{-3} which has 3 significant figures (ignore the leading zero but include the one 'inside' the number), so our answer should be given to 3 significant figures. Rounding up the third of these, leads to a final value for the volume of 24.8 cm^3.

**1.3
Uncertainty in
measurement**

So far, we have addressed the question of how many figures are justified when quoting the numerical value of a particular quantity. In other words, we have seen how to specify the appropriate precision of a physical quantity. We now need to consider how we can judge the accuracy of our value.

The rate constant for the hydrolysis of methyl acetate is $1.32 \times 10^{-4}\,\text{s}^{-1}$ at 25°C; the dissociation constant of hydrofluoric acid is 5.62×10^{-4}; the enthalpy of formation of methane from its elements is $-74.9\,\text{kJ mol}^{-1}$.

The object of many physical chemistry experiments is to obtain the value of some numerical quantity. Examples are the rate constant for a reaction, the dissociation constant of an acid or the enthalpy change accompanying a reaction. Frequently, in undergraduate work, the accepted values for these quantities are well documented and a student can assess the uncertainty in his or her determination of a particular value by consulting the literature.

In genuine research work, this is not the case. You might be studying a new reaction or using a new technique, and have no idea what the exact value is going to be. In this case, it is vital to be able to estimate the uncertainty associated with such a value. Even if the value is known, you need to be able to state whether your experimental value agrees with the literature value within the bounds of experimental error.

An alternative treatment of uncertainty analysis is given in *Journal of Chemical Education*, **61**, 780–1, 1984.

There are two approaches to the determination of experimental uncertainties, or errors. When carrying out research, the determination of a physical quantity would be repeated many times, and statistical methods can then be used as described later in this section. You are more likely to be in the position of having a fixed number of hours to perform an experiment, after which the demands of the course dictate that you move on to new work. The results of the class may be pooled to allow a statistical treatment but, more frequently, you will be left with a single determination of your value on which you need to estimate the uncertainty. In this case, you will have to estimate the maximum probable error.

1.3.1   Determining the maximum probable error

This is relatively straightforward when applied to the reading of a single instrument, requiring only a commonsense approach. You may decide that you can read a burette to $\pm 0.05\,\text{cm}^3$, a balance to ± 0.0001 g or a digital voltmeter to ± 0.001 V for example. We use the $\pm$ symbol in the specification of errors to show that they have an equal chance of being too high or too low by this amount.

The Greek letter Δ (capital delta) is conventionally used to denote a difference in two values of a quantity. It is much used in physical chemistry, and we will meet it again in Chapter 2.

You need to be able to distinguish between three definitions of errors. To illustrate these, suppose you have estimated that the uncertainty on a quantity X is ΔX:

- the **absolute error** is simply ΔX
- the **fractional error** is $\Delta X / X$
- the **percentage error** is $100 \times \Delta X / X$

Worked example 1.5

An actual volume of $25.00\,\text{cm}^3$ is measured as $25.15\,\text{cm}^3$. What is (a) the absolute error, (b) the fractional error and (c) the percentage error in this quantity?

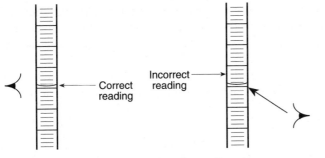

Figure 1.5 Effect of parallax when reading a burette.

Chemical background

The uncertainty involved in this example is rather more than you should hope to achieve in practice! It helps to ensure that the burette is mounted vertically, and that the effects of parallax are avoided when readings are taken, as shown in Figure 1.5.

Solution to worked example

(a) The absolute error is ±0.15 cm^3; we use the $\pm$ sign because we would not generally know in which direction the error was.
(b) The fractional error is

$$\pm\frac{0.15 \text{ cm}^3}{25.00 \text{ cm}^3} = \pm0.006$$

(c) The percentage error will be 100 times the fractional error, which is $\pm0.6\%$.

1.3.2 Combining errors

The determination of the value of a physical quantity in an experiment will usually involve several measurements followed by one or more calculations involving these quantities. As these are combined using mathematical operations such as addition, subtraction, multiplication and division, the corresponding uncertainties are also combined using a set of rules. Let us suppose that we have measured two quantities X and Y and have estimated the uncertainties on them to be ΔX and ΔY respectively. The errors are combined as follows:

(a) If X and Y are added, or one is subtracted from the other, the uncertainty on the result will be

$$\sqrt{(\Delta X)^2 + (\Delta Y)^2}$$

(b) If X and Y are multiplied, or one is divided by the other, the fractional error on the result will be

$$\sqrt{\left(\frac{\Delta X}{X}\right)^2 + \left(\frac{\Delta Y}{Y}\right)^2}$$

Following on from this, if X is raised to the power n, the fractional uncertainty on the product will be

$$\frac{\Delta X}{X}\sqrt{n}$$

Worked example 1.6

The initial and final burette readings in a titration are (5.00 ± 0.05) cm^3 and (31.85 ± 0.05) cm^3 respectively. What is the uncertainty in the titre?

Chemical background

While it is probably more usual to refill a burette to the zero mark before starting each titration, there is an advantage to not doing this. Using a greater length of the burette will help to average out variations in the bore. It is important, however, to ensure that you do not run out of solution before the end point has been reached!

The capacity tolerances of precision volumetric glassware are specified by the National Physical Laboratory as 'Grade A' or 'Grade B'. It should be possible to achieve an accuracy of within 0.05% with a pipette, and within 0.1% with a burette.

Solution to worked example

The value of the titre is found by subtracting the initial volume from the final volume, so we have

$$\text{titre} = (31.85 - 5.00) \text{ cm}^3 = 26.85 \text{ cm}^3$$

We therefore need to combine the absolute errors as given above, which leads to

$$\text{uncertainty} = \sqrt{(0.05 \text{ cm}^3)^2 + (0.05 \text{ cm}^3)^2}$$
$$\approx 0.07 \text{ cm}^3$$

It is worth noting that this value has been obtained independently of the titre value, and so is valid for *any* titration in which the burette has been consistently read to ± 0.05 cm^3. We would express the final titre as (26.85 ± 0.07) cm^3.

The brackets are used here to show that the units apply to both the titre value and its uncertainty but, by convention, they may be omitted in which case the titre would be given as $26.85 \pm 0.07 \text{ cm}^3$.

Worked example 1.7

Calculate the solubility of magnesium ethanedioate from the following information. A volume $14.85 \pm 0.07 \text{ cm}^3$ of acidified potassium permanganate solution of concentration $0.0050 \pm 0.000\,05 \text{ mol dm}^{-3}$ is required to oxidize $20.0 \pm 0.1 \text{ cm}^3$ of a saturated solution of the oxalate. This reaction can be represented by the equation

$$5C_2O_4^{2-} + 2MnO_4^- + 16H^+ \rightarrow 10CO_2 + 8H_2O + 2Mn^{2+}$$

Chemical background

This is an example of a redox titration in which electron transfer is involved. The permanganate or manganate (VII) ion MnO_4^- is reduced to manganese (II) ions

$$MnO_4^- + 8H^+ + 5e^- \rightarrow Mn^{2+} + 4H_2O$$

while the ethanedioate ion $C_2O_4^{2-}$ is oxidized to carbon dioxide

$$C_2O_4^{2-} \rightarrow 2CO_2 + 2e^-$$

Solution to worked example

The only information which needs to be extracted from this complicated equation is that the ratio amount of ethanedioate : amount of permanganate is $5:2$. This gives us the equation

$$\frac{\text{amount of ethanedioate}}{\text{amount of permanganate}} = \frac{5}{2}$$

Substituting values for the concentration and volume leads to

$$\frac{20.0 \text{ cm}^3 \times [C_2O_4^{2-}]}{14.85 \text{ cm}^3 \times 0.0050 \text{ mol dm}^{-3}} = \frac{5}{2}$$

which can be rearranged to give

$$[C_2O_4^{2-}] = \left(\frac{5}{2}\right) \times \frac{14.85 \text{ cm}^3 \times 0.0050 \text{ mol dm}^{-3}}{20.0 \text{ cm}^3}$$

Use of a calculator gives the value 9.3×10^{-3} mol dm^{-3}, applying the rules we have learned for significant figures and realizing that the trailing zero in the value of the permanganate concentration is significant.

If we now look at the equation we used to evaluate $[C_2O_4^{2-}]$, we see that the only mathematical operations involved are multiplication and division. Therefore we can evaluate the overall uncertainty by using the rule for fractional errors given above. First, we need to calculate the fractional error on each quantity.

Fractional error on titre

$$= \pm \frac{0.07 \text{ cm}^3}{14.85 \text{ cm}^3} \simeq \pm 0.0047$$

Fractional error on permanganate concentration

$$= \pm \frac{0.000\,05 \text{ mol dm}^{-3}}{0.0050 \text{ mol dm}^{-3}} = \pm 0.0100$$

Fractional error on ethanedioate volume

$$= \pm \frac{0.1 \text{ cm}^3}{20.0 \text{ cm}^3} = \pm 0.005$$

The stoichiometric ratio, $\frac{5}{2}$, is an exact number so the associated uncertainty is effectively zero.

The overall fractional error is then

$$\sqrt{0.0047^2 + 0.0100^2 + 0.005^2} = \sqrt{0.000147} \simeq 0.0121$$

We multiply this by our concentration value to give the absolute error:

$$\text{absolute error} = 0.0121 \times 9.3 \times 10^{-3} \text{ mol dm}^{-3}$$

$$\simeq 0.11 \times 10^{-3} \text{ mol dm}^{-3}$$

The final value for the concentration is then given as $(9.3 \pm 0.1) \times 10^{-3}$ mol dm^{-3}. When scientific notation is involved as here, it is usually clearer to quote the uncertainty as a number to the same power of 10 as the value to which it refers, even if it means that the uncertainty is not given as a number between 1 and 10.

Worked example 1.8

The concentrations of hydrogen, iodine and hydrogen iodide in the gaseous equilibrium

$$H_2 + I_2 \rightleftharpoons 2HI$$

are related by the equilibrium constant K, so that if all other quantities are known the hydrogen concentration can be determined from the equation

$$[H_2] = \frac{[HI]^2}{K[I_2]}$$

If the equilibrium constant K has been determined as 46.0 ± 0.2, calculate the concentration of hydrogen, with its associated uncertainty, when $[HI] = (17.2 \pm 0.2) \times 10^{-3} \, mol \, dm^{-3}$ and $[I_2] = (2.9 \pm 0.1) \times 10^{-3} \, mol \, dm^{-3}$.

More data on this equilibrium reaction will be found in *Journal of the American Chemical Society*, **63**, 1377, 1941.

Chemical background

This is an example of a chemical equilibrium which has been studied extensively. Although we have dealt with this in terms of concentrations, it is also possible to do so in terms of the partial pressures of each component. No matter what the starting concentrations or pressures, at equilibrium, the relationship given will always be obeyed at a given temperature; at a different temperature, we would have to use a different value for the equilibrium constant K. One consequence of this relationship is that it does not matter whether we start with a mixture of hydrogen or iodine, or by allowing hydrogen iodide to decompose. The same equilibrium will be reached in each case.

One practical way in which this equilibrium could be studied would be to heat hydrogen iodide in a bulb of known volume as shown in Figure 1.6. Once equilibrium has been reached, this could then be cooled rapidly to room temperature to stop or 'quench' the reaction. The concentration of iodine can then be determined by titration with sodium thiosulphate.

Solution to worked example

Using the equation given to evaluate the concentration of hydrogen gives us

$$[H_2] = \frac{(17.2 \times 10^{-3} \, mol \, dm^{-3})^2}{46.0 \times 2.9 \times 10^{-3} \, mol \, dm^{-3}}$$

which can be evaluated using a calculator to give the value of $2.218 \times 10^{-3} \, mol \, dm^{-3}$. Since we are going to evaluate an error limit from the information

Heat

Figure 1.6 Measuring the equilibrium constant for the dissociation of hydrogen iodide.

given there is no point at this stage worrying too much about the number of significant figures to quote. It is best to err on the side of caution, however, to make sure that we include enough significant figures to allow rounding at a later stage.

Notice that, since only multiplication and division are involved in the equation given, we need to combine the fractional error on each quantity. Since the equation contains $[HI]^2$ rather than $[HI]$, the first thing we will do is to evaluate the fractional error on $[HI]^2$.

$$\frac{0.2}{17.2} = 1.2 \times 10^{-2}$$

Since the concentrations and their associated errors are multiplied by the same powers of 10, these can be ignored when we are dealing with fractional errors. Using the equation above, we have

$$\text{Fractional error on } [HI]^2 = \sqrt{2} \times \text{fractional error on } [HI]$$

$$= \sqrt{2} \times 1.2 \times 10^{-2}$$

$$\simeq 1.7 \times 10^{-2}$$

The remaining fractional errors are straightforward to calculate:

$$\text{Fractional error on } K = \frac{0.2}{46.0} \simeq 4.3 \times 10^{-3}$$

$$\text{Fractional error on } [I_2] = \frac{0.1}{2.9} \simeq 3.4 \times 10^{-2}$$

We now combine these three fractional errors using the rule for products and quotients:

$$\text{Fractional error on } [H_2]$$
$$= \sqrt{(1.7 \times 10^{-2})^2 + (4.3 \times 10^{-3})^2 + (3.4 \times 10^{-2})^2}$$

$$\simeq 0.0382$$

To obtain the absolute error on $[H_2]$ we need to multiply this fractional error by the value of $[H_2]$ itself.

$$\text{Absolute error on } [H_2] = 0.0382 \times 2.218 \times 10^{-3} \text{ mol dm}^{-3}$$

$$= 0.0847276 \times 10^{-3} \text{ mol dm}^{-3}$$

$$\simeq 8.47 \times 10^{-5} \text{ mol dm}^{-3}$$

We can now write our final expression for the hydrogen concentration as

$$[H_2] = (2.22 \pm 0.08) \times 10^{-3} \text{ mol dm}^{-3}$$

1.3.3   Statistical treatment of errors

Statistics is a discipline itself, so we will do no more than look at the basics here to enable us to assign reasonable error limits. As noted earlier, the statistical treatment is only of use when we have repeated measurements of the same quantity. At this level, it is only important to understand a few basic definitions of statistical quantities and to be able to apply them to sets of data.

To illustrate these definitions, consider the small data set

$$1.2 \quad 1.4 \quad 1.2 \quad 1.5 \quad 1.3$$

which contains five numbers. One way of representing these is by using the notation x_i, where i is an integer so that x_1 is the first number (1.2), x_2 is the second number (1.4), and so on.

The **arithmetic mean** is obtained by adding all the values of the quantity together and then dividing by the number of measurements taken. It can be denoted as

$$\bar{x} = \sum \frac{x_i}{n}$$

where n is the number of numbers.

Do not be put off by the use of some strange looking mathematical symbols here. This is a good place to introduce some terminology as the concepts are still simple, and it will make life easier with some of the other quantities. The symbol $\sum$ (a capital Greek sigma) simply means 'add up'. So if we add the five numbers given above we get 6.6, which is equal to $\sum x_i$. We then divide this by n, the number of values which, in this case, is 5. This gives us a value for the arithmetic mean of

$$\frac{6.6}{5} = 1.32 \simeq 1.3$$

The **median** is obtained by arranging all the values obtained in ascending (or descending) order; if there are an odd number of values the median is the middle one, otherwise it is the arithmetic mean of the two middle values. So, arranging our data set in ascending order (lowest first) gives us

$$1.2 \quad 1.2 \quad 1.3 \quad 1.4 \quad 1.5$$

As we have an odd number of values, the central one is the third, and so the value of the median is 1.3.

The **mode** is the most frequently occurring value. Looking at our example, only 1.2 appears more than once, so this is the mode in this case.

The arithmetic mean, median and mode can all be thought of as different kinds of 'average'. What none of these quantities tells us, however, is how the values are spread about these averages. To do this, we need to consider the variance and standard deviation.

The **variance** σ^2 is the average of the squares of the deviations from the true value of the quantity being measured. However, since we are taking our estimate of this value as our arithmetic mean, we need to divide by $n - 1$ instead of n, where n is the number of values we have. We could only divide by n if we were sure that our arithmetic mean was equal to the true mean; normally it will only be an estimate. Using the terminology introduced above,

$$\sigma^2 = \sum \frac{(x_i - \bar{x})^2}{n - 1}$$

This means that we take each value of x_i in turn, subtract the arithmetic mean, square the result and divide by $n - 1$. We then add each of these terms, the total sum being the variance. Its square root is the **standard deviation** σ, which is a measure of the spread of values about the mean. Notice that the standard deviation is of more use than the variance because it has the same units as the quantity being measured. We would expect 95% of all values to lie within two standard deviations of the mean, so that two standard deviations can be taken as a reasonable error estimate for each individual reading.

We still need to address the problem of how closely our mean value is a measure of the true value. The best estimate would be obtained by taking several samples of size n; the standard error is then defined as

$$\frac{\sigma}{\sqrt{n}}$$

where σ has been determined as above. This expression is obtained by considering the variance of the sample mean, which is σ^2/n; the square root of this is taken because it has the same units as the measured quantity.

Doubling the standard error gives us a **confidence limit** of 95% on the mean; in other words, if we performed many determinations of the mean, we would expect 95% of the results to lie within this limit.

From the equations given for the standard deviation and the standard error of the mean, you should be able to see that the mean will have a lower estimated error than each individual measurement, as we might expect.

Worked example 1.9

Ten determinations of the dipole moment of hydrogen chloride gas gave the following values (in units of D):

$$1.048 \quad 1.047 \quad 1.053 \quad 1.048 \quad 1.051$$

$$1.053 \quad 1.045 \quad 1.051 \quad 1.047 \quad 1.047$$

Calculate (a) the arithmetic mean, (b) the median, (c) the mode, (d) the variance and (e) the standard deviation of these values.

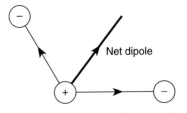

Figure 1.7   Net dipole resulting from two polar bonds.

Chemical background

When placed in an electric field, all molecules have an induced dipole moment, since the centres of gravity of positive and negative charges do not coincide. Polar molecules also have a permanent dipole moment (Figure 1.7) which can be determined from measurements of the dielectric constant, also called the relative permittivity. In the case of hydrogen chloride, this can be done using a specially constructed gas capacitance cell.

Solution to worked example

(a) The sum of the values is 10.49 D, so the arithmetic mean is

$$\frac{10.49 \text{ D}}{10} = 1.049 \text{ D}$$

Notice that by using the rules of significant figures (with 10 being an exact integer) we keep the same number of decimal places as in the original data.

(b) Rearranging the data in ascending order gives

$$1.045 \quad 1.047 \quad 1.047 \quad 1.047 \quad 1.048$$
$$1.048 \quad 1.051 \quad 1.051 \quad 1.053 \quad 1.053$$

Since there are an even number of data we need to look at the fifth and sixth values; these are actually the same so their arithmetic mean and the median of the data is 1.048 D.

(c) By inspection, the value of 1.047 D appears three times and so this is the mode.

(d) To calculate the variance, we need to calculate the square of the deviation from the mean for each value which occurs:

$$\sigma^2 = \left(\frac{1}{10-1}\right) \times [1 \times (1.045 - 1.049)^2$$
$$+ 3 \times (1.047 - 1.049)^2$$
$$+ 2 \times (1.048 - 1.049)^2$$
$$+ 2 \times (1.051 - 1.049)^2$$
$$+ 2 \times (1.053 - 1.049)^2] \text{ D}^2$$

$$\sigma^2 = \left(\frac{1}{9}\right)[1.6 + 1.2 + 0.2 + 0.8 + 3.2] \times 10^{-5}\,\mathrm{D}^2$$

$$= 7.8 \times 10^{-6}\,\mathrm{D}^2$$

(e) The standard deviation is simply the square root of this value, which is 0.003 D.

1.3.4 Statistics using a calculator

While it is useful to have a knowledge of the underlying equations used, most of the simple statistics used in chemistry can be performed using the functions on a scientific calculator ▦. Figure 1.8 shows the series of steps required to enter a sample data set and determine the arithmetic mean and standard deviation. As all types of calculator are slightly different however, it is worth consulting your manual so that you know how to:

1. enter statistics mode
2. enter data items
3. obtain output.

Worked example 1.10

The following determinations of the internuclear distance in hydrogen chloride were made (values in Å):

$$1.279 \quad 1.277 \quad 1.277 \quad 1.281$$

$$1.278 \quad 1.283 \quad 1.281 \quad 1.280$$

Use statistical methods to obtain an estimate of the H—Cl bond length with its uncertainty.

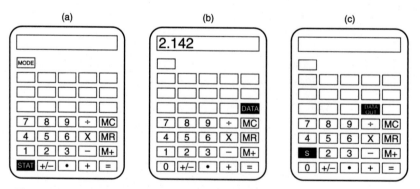

Figure 1.8 Procedure for performing statistical calculations using a calculator: (a) to enter statistics mode; (b) to enter data items; and (c) to obtain output.

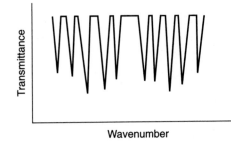

Figure 1.9   The vibration–rotation spectrum of hydrogen chloride gas.

Chemical background

The H—Cl bond length can be determined by measuring the vibrational–rotational spectrum of hydrogen chloride gas, a part of which is shown in Figure 1.9. This gives the value of the rotational constant B, which is defined by

$$B = \frac{h}{8\pi^2 I}$$

where h is Planck's constant and I is the moment of inertia defined by

$$I = \mu r^2$$

where μ is the reduced mass and r the internuclear distance.

The ångström (Å) is a non-SI unit which still finds much favour among chemists working on structural problems. It is defined as 10^{-10} m, and consequently a single bond between two carbon atoms has a length of 1.54 Å. The use of numbers around 1 and 2 seems to be preferred to the fractional nm (e.g. C—C 0.154 nm) or larger pm (C—C 154 pm).

Solution to worked example

We need to calculate the standard deviation of these data. The quickest way to do this is using a calculator , but here the manual working is also shown. We begin by calculating the arithmetic mean.

$$\text{Sum of data} = 10.236 \text{ Å}$$

$$\text{Arithmetic mean} = \frac{10.236 \text{ Å}}{8}$$

$$\simeq 1.280 \text{ Å}$$

The next stage is to obtain the variance.

$$\sigma^2 = \left(\frac{1}{8-1}\right) \times [2 \times (1.277 - 1.280)^2$$

$$+ 1 \times (1.278 - 1.280)^2$$

$$+ 1 \times (1.279 - 1.280)^2$$

$$+ 1 \times (1.280 - 1.280)^2$$

$$+ 2 \times (1.281 - 1.280)^2$$

$$+ 1 \times (1.283 - 1.280)^2]$$

$$= \left(\frac{1}{7}\right)[1.8 + 0.4 + 0.1 + 0.0 + 0.2 + 0.9] \times 10^{-6} \text{ Å}^2$$

$$\approx 4.86 \times 10^{-6} \text{ Å}^2$$

The standard deviation is the square root of this, 0.002 Å. It then follows that the standard error will be

$$\frac{\sigma}{n} = \frac{0.002 \text{ Å}}{\sqrt{8}}$$

$$\approx \frac{0.002 \text{ Å}}{2.828}$$

$$\approx 7 \times 10^{-4} \text{ Å}$$

If we take twice the standard error as a reasonable error estimate, we have the H$-$Cl bond length as (1.280 ± 0.001) Å.

Exercises

1. Write the following numbers to 2 significant figures:
   (a) 348   (b) 0.0012   (c) 0.5437   (d) 12.65
2. Write the following numbers to 2 decimal places:
   (a) 13.847   (b) 0.2549   (c) 99.543 21   (d) 0.007 54
3. Write the following numbers to 3 significant figures:
   (a) 78 521   (b) 0.006 7543   (c) 12.005 485   (d) 80 006
4. Write the following numbers to 1 decimal place:
   (a) 3.098   (b) 12.76   (c) 0.000 62   (d) 1.049
5. Give the results of the following calculations to the appropriate number of figures:

   (a) $12.36 + 3.2$ (b) $31.8 - 12.72$ (c) 1.9×1.235 (d) $\dfrac{4.673}{2.1}$

   (e) $\dfrac{(34.2 - 1.75)}{2.34}$

6. Give the results of the following calculations to the appropriate number of significant figures:

(a) 2×32.74 (b) $3 \times 12.01 \times 32.1$ (c) $\left(\dfrac{5}{2}\right) \times \left(\dfrac{3.07}{6.18}\right)$

7. The side of a cuboid are measured as 21.0 ± 0.1 cm, 16.0 ± 0.1 cm and 32.0 ± 0.2 cm. What is (a) the error, (b) the fractional error and (c) the percentage error in the volume of the cuboid?

8. A car covered a distance of 9.2 ± 0.1 km in 295 ± 1 s. Use the formula

$$\text{speed} = \frac{\text{distance}}{\text{time}}$$

to calculate the average speed of the car and its associated uncertainty.

9. Calculate (a) the arithmetic mean, (b) the median and (c) the mode of the following numbers:

    10.65   10.62   10.58   10.67   10.66   10.61   10.61

10. Calculate the standard deviation of the following numbers:

    23.72   24.01   23.86   24.09   23.68   23.99

Problems

1. Write the values of each of these quantities to (i) 3 significant figures and (ii) 2 decimal places:
(a) a bond length of 1.542 Å
(b) a temperature of 25.01°C
(c) an enthalpy change of -432.876 kJ mol^{-1}
(d) the relative atomic mass of oxygen, 15.9994
(e) the ideal gas constant, 8.314 J K^{-1} mol^{-1}

2. Give the results of the following calculations to the appropriate number of figures:

(a) $\text{density} = \dfrac{\text{mass}}{\text{volume}}$

$= \dfrac{17.098 \text{ g}}{6.2 \text{ cm}^3}$

(b) mass of sample $= 20.1452$ g $- 5.12$ g

(c) internal energy $\Delta U = Q + W$

$= 321.4$ kJ mol^{-1} $- 162$ kJ mol^{-1}

(d) $pV = nRT$

$= 3.2$ mol $\times 8.314$ J K^{-1} mol^{-1} $\times 298$ K

(e) equilibrium constant $K_c = \dfrac{0.106 \text{ mol dm}^{-3} \times 0.098 \text{ mol dm}^{-3}}{0.250 \text{ mol dm}^{-3}}$

3. The overall entropy change for any process is equal to the sum of the entropy changes for the individual steps involved. In one experiment, these changes were measured as $3.90 \pm 0.05 \text{ J K}^{-1} \text{ mol}^{-1}$, $38.4 \pm 0.1 \text{ J K}^{-1} \text{ mol}^{-1}$ and $5.05 \pm 0.05 \text{ J K}^{-1} \text{ mol}^{-1}$. Determine the uncertainty in the sum of these three entropy changes.

4. The volume V of an ideal gas can be found by rearranging the ideal gas equation, and is given by

$$V = \frac{nRT}{p}$$

where n is the amount of gas, R the ideal gas constant ($8.314 \text{ J K}^{-1} \text{ mol}^{-1}$), T the absolute temperature and p the pressure. Calculate the volume, with its associated uncertainty, for $2.35 \pm 0.1 \text{ mol}$ of gas at a temperature of $298.4 \pm 0.1 \text{ K}$ and a pressure of $1.033 \pm 0.005 \text{ kPa}$.

5. Calculate the arithmetic mean, median, mode, variance and standard deviation of the following values of the ionization energy of sodium (in kJ mol^{-1}):

$$495 \quad 497 \quad 493 \quad 493 \quad 496 \quad 492$$

Thermodynamics 2

Thermodynamics is one of the branches of physical chemistry which students often find most confusing. It involves many different concepts, each of which seems to require a different mathematical approach and a different terminology. However, it is also a very important subject, whose ideas underpin the whole of chemistry and without which a full appreciation of chemical concepts is not possible.

In this chapter, you will be taken through some of the main areas of thermodynamics and shown the maths required to deal with each of them. We will cover a number of different areas of mathematics, including possibly the first experience you have had of calculus. The terminology used can sometimes be offputting, but each new symbol will be explained as we meet it. You will find the maths easier if you can remember the meanings of the symbols introduced.

2.1 The Born–Haber cycle

One of the fundamental ideas in thermodynamics is that the change in value of certain variables is equal to the sum of other smaller changes which give the same overall result. For example, when a chemical reaction involving some kind of molecular rearrangement takes place, the change in a variable called the enthalpy is the same as if we had dissociated all the reactants into atoms and then allowed them to recombine into the required products. Of course, we know that this does not really happen, but it is a useful concept which allows us to calculate quantities which are difficult to measure experimentally. An analogy from everyday life is that if we climb a mountain, the height we reach is independent of the path we choose to take.

Some of the thermodynamic variables for which this is true are those called the enthalpy, entropy, Gibbs free energy and chemical potential; these are known as **state functions**. The two most common non-state functions are work and heat. You will meet all of these variables at some point in your study of physical chemistry.

The mathematical consequence of using state functions is that we can calculate the value of a change in a particular function by adding together other changes which refer to a different route, even though the starting and end points are the same. While this is very useful to chemists in being able to calculate enthalpy changes for reactions where they cannot be measured, there is an important point to remember. We saw in the last chapter that any experimental measurement is subject to an uncertainty, and if we are adding together several experimental values we are likely to obtain an overall uncertainty which is much higher. It follows that an enthalpy change calculated in this way is likely to be less certain than one which has been measured directly.

2.1.1   Combining errors

The mathematical background to this has been discussed in section 1.3.

Worked example 2.1

The symbol MX is used to denote a generalized formula, where M represents a metal cation and X an anion. Examples would be CsCl and KF. These are $1:1$ compounds where there are equal numbers of cations and anions. A $1:2$ structure would have the generalized formula MX_2, examples being CaF_2 and MnS_2.

The standard lattice enthalpy $\Delta_{\text{lattice}}H^\ominus$ of a $1:1$ ionic crystal MX is given by the equation

$$\Delta_{\text{lattice}}H^\ominus = -\Delta_f H^\ominus + \Delta_{\text{sub}}H^\ominus + \tfrac{1}{2}\Delta_{\text{at}}H^\ominus + \Delta_{\text{ion}}H^\ominus + \Delta_{\text{ea}}H^\ominus$$

where $\Delta_f H^\ominus$ is the standard enthalpy of formation of MX, $\Delta_{\text{sub}}H^\ominus$ is the standard enthalpy of sublimation of M, $\Delta_{\text{at}}H^\ominus$ is the standard enthalpy of atomization of X_2, $\Delta_{\text{ion}}H^\ominus$ is the ionization enthalpy of M and $\Delta_{\text{ea}}H^\ominus$ is the electron affinity of X.

Calculate the electron affinity of Cl, together with its associated uncertainty, using the following data:

Standard enthalpy of formation of $NaCl_{(s)}$ = $-411 \pm 2 \text{ kJ mol}^{-1}$
Enthalpy of sublimation of Na = $109 \pm 1 \text{ kJ mol}^{-1}$
Enthalpy of atomization of $Cl_{2(g)}$ = $242 \pm 1 \text{ kJ mol}^{-1}$
Ionization enthalpy of $Na_{(g)}$ = $494 \pm 2 \text{ kJ mol}^{-1}$
Standard lattice enthalpy of NaCl = $760 \pm 3 \text{ kJ mol}^{-1}$

Chemical background

A consideration of the lattice enthalpy finds use when we are interested in the forces and energies within ionic crystals, as it is possible to compare the value with that obtained from theoretical calculations. In contrast to the problem here, we would normally calculate the value of the lattice enthalpy using experimental values of the other enthalpy changes; a direct experimental determination is not possible. The lattice enthalpy in this problem actually refers to the enthalpy change accompanying the reaction

$$NaCl_{(s)} \rightarrow Na^+_{(g)} + Cl^-_{(g)}$$

This contrasts with the enthalpy of formation, which is the enthalpy for the reaction

$$Na_{(s)} + \tfrac{1}{2}Cl_{2(g)} \rightarrow NaCl_{(s)}$$

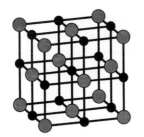

Figure 2.1 The crystal lattice of sodium chloride.

The other enthalpy changes required are the enthalpy of sublimation:

$$Na_{(s)} \rightarrow Na_{(g)}$$

and the enthalpy of ionization

$$Na_{(g)} \rightarrow Na^+_{\ (g)} + e^-$$

For anions, we can define the electron affinity, which in this case is the enthalpy change for the reaction

$$Cl^-_{\ (g)} \rightarrow Cl_{(g)} + e^-$$

This reaction is the reverse of what you might expect, and it is often written the other way round with accompanying phrases used to obtain the correct sign for the enthalpy change.

Sodium chloride exists as a regular cubic lattice, shown in Figure 2.1.

Solution to worked example

The calculation of $\Delta_{ea}H^\ominus$ is straightforward, and simply involves rearranging the given equation to leave this term as the subject. This gives

$$\Delta_{ea}H^\ominus = \Delta_{lattice}H^\ominus + \Delta_f H^\ominus - \Delta_{sub}H^\ominus - \tfrac{1}{2}\Delta_{at}H^\ominus - \Delta_{ion}H^\ominus$$

$$= [760 - 411 - 109 - (0.5 \times 242) - 494] \text{ kJ mol}^{-1}$$

$$= -375 \text{ kJ mol}^{-1}$$

We can now turn our attention to determining the uncertainty in this quantity. Look first of all at the enthalpy of atomization of $Cl_{2(g)}$. This has been multiplied by 0.5, which we can take as being an exact constant with zero error. Therefore the fractional error on $0.5\Delta_{at}H^\ominus$ will be given by

$$\sqrt{\left(\frac{0}{0.5}\right)^2 + \left(\frac{1}{242}\right)^2} = \frac{1}{242}$$

the same as the fractional error on $\Delta_{at}H^\ominus$. The absolute error in this quantity will therefore be

$$\left(\frac{1}{242}\right) \times 0.5 \times 242 \text{ kJ mol}^{-1} = 0.5 \text{ kJ mol}^{-1}$$

Note that the overall error is, inevitably, greater than any of the individual uncertainties used in the calculation. For this reason, a direct experimental measurement may be preferable to a calculated value such as this.

Having done this, we have a set of uncertainties on five values which are combined by means of addition and subtraction. The absolute error is then given by

$$\sqrt{[(3)^2 + (2)^2 + (1)^2 + (0.5)^2 + (2)^2]} \text{ kJ mol}^{-1} = \sqrt{18.25} \text{ kJ mol}^{-1}$$

$$\simeq 4.3 \text{ kJ mol}^{-1}$$

The value of $\Delta_{ea}H^{\ominus}$ is now $-375 \pm 4 \text{ kJ mol}^{-1}$.

2.2
Heat capacity

The branch of thermodynamics concerned with heat changes is **thermochemistry**. The principal property which governs the thermal behaviour of a substance is the heat capacity, which can be defined at a simple level as the heat required to raise the temperature of the substance by 1 K. More precisely, we can identify two values of the heat capacity according to the conditions under which any change is brought about. They are defined as

$$C_p = \text{heat capacity at constant pressure}$$

and

$$C_v = \text{heat capacity at constant volume}$$

2.2.1 Expansion of brackets

Mathematical expressions can be made more concise by the use of brackets. This is generally useful for giving general expressions, but there are times when it is desirable to expand such an expression into its individual terms. For example, if we have an expression such as

$$a(b + c)$$

this is expanded by multiplying the quantity outside the bracket (in this case, a) with each term inside the bracket.

$$a(b + c) = ab + ac$$

If two brackets are multiplied together, such as

$$(a + b)(c + d)$$

then each term in the first bracket needs to be multiplied with each term in the second.

$$(a + b)(c + d) = a(c + d) + b(c + d)$$

Multiplying out again gives

$$(a + b)(c + d) = a(c + d) + b(c + d)$$
$$= ac + ad + bc + bd$$

It may, of course, be possible to simplify the resulting expression in many cases.

Worked example 2.2

The value of the enthalpy change ΔH is given in terms of the heat capacity C_p and the temperature change ΔT as

$$\Delta H = C_p \, \Delta T$$

Write an expression for the absolute enthalpy H_2 at temperature T_2 in terms of the absolute enthalpy H_1 at temperature T_1 and the heat capacity.

Chemical background

This relationship arises directly from the definition of **enthalpy**, which is the quantity of heat supplied to a system at constant pressure. It can be calculated from the formula

$$H = U + pV$$

where U is the internal energy, p is the pressure and V the volume.

The expression given in the problem is only valid when C_p can be considered to be independent of temperature. In the next Worked Example, we will see an example of C_p depending on temperature.

Solution to worked example

We have already met the use of the Δ notation when discussing enthalpy changes in Worked Example 2.1. While a quantity such as the enthalpy H does have absolute values, in thermodynamics we are frequently more interested in changes to these values. We use the notation ΔH to denote such an enthalpy change, where this is defined as

$$\Delta H = H_2 - H_1$$

The convention is always to subtract the initial value (in this case H_1) from the final value (H_2). Similarly, the change in temperature is given by

$$\Delta T = T_2 - T_1$$

and so the expression becomes

$$H_2 - H_1 = C_p(T_2 - T_1)$$

We now expand the right-hand side of this equation by multiplying the terms in the bracket by C_p to give

$$H_2 - H_1 = C_p T_2 - C_p T_1$$

We were asked to obtain an expression for H_2, so we rearrange the equation, making H_2 the subject, by adding the term H_1 to both sides.

$$H_2 = H_1 + C_p T_2 - C_p T_1$$

This is a simple problem, yet being able to perform manipulations like this is vital to a mastery of thermodynamics.

2.2.2   Polynomial expressions

You have probably seen expressions like

$$5x^3 + 2x^2 + 3x + 10$$

which is an example of a **polynomial expression in** x. The characteristic of a polynomial is that it consists of a series of terms consisting of the variable (in this case x) raised to an integral power. The highest value of these powers is called the **order** of the expression, and the constants which multiply each of these terms are called **coefficients**. So in the example above, the highest power is 3 so this is a polynomial of order 3. We also note that

- the coefficient of x^3 is 5
- the coefficient of x^2 is 2
- the coefficient of x is 3.

The final value of 10 is called the **constant term**. Note that any of these terms can be missing, so that the expression $2x^4 + 3x$ is also an example of a polynomial expression, although it does not contain terms in x^3 or x^2, nor a constant.

Special names are given to polynomials with the lowest orders. These are:

- **linear** for an expression of order 1
- **quadratic** for an expression of order 2, and
- **cubic** for an expression of order 3.

Examples of each of these are shown in Figure 2.2.

Worked example 2.3

The heat capacities C_p for the species involved in the reaction

$$H_{2(g)} + \tfrac{1}{2}O_{2(g)} \rightarrow H_2O_{(g)}$$

can be expressed as a function of temperature T as a polynomial

$$C_p = a + bT + cT^2$$

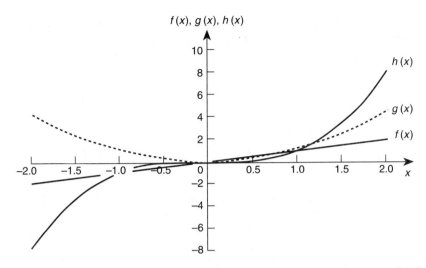

Figure 2.2   Graphs of representative linear, quadratic and cubic functions, $f(x)$, $g(x)$ and $h(x)$, respectively.

where a, b and c are constants. Use the values in the table below to obtain an expression for the change in heat capacity as a function of temperature when this reaction takes place.

	$\dfrac{a}{\text{J K}^{-1}\,\text{mol}^{-1}}$	$\dfrac{b}{10^3\,\text{J K}^{-2}\,\text{mol}^{-1}}$	$\dfrac{c}{10^7\,\text{J K}^{-3}\,\text{mol}^{-1}}$
$H_{2(g)}$	29.07	−0.836	20.1
$O_{2(g)}$	25.72	12.98	−38.6
$H_2O_{(g)}$	30.36	9.61	11.8

Chemical background

The calculation of the expression is an important intermediate step when determining how an enthalpy change for a given reaction varies with temperature. As well as the form of polynomial here, an expression such as

$$C_p = a + bT + \frac{c}{T^2}$$

can be used. Note that this is not actually a polynomial because of the final term, but it can be handled in exactly the same way as the expression in this problem.

Such expressions are able to fit experimental data to within about 0.5% over a wide range of temperatures.

The chemical industry uses large quantities of water, the majority for removing heat. Pure water may be required for some of the other chemical uses, and this can be produced by distillation.

In fact, a mixture of hydrogen and oxygen gases will only react to give steam (gaseous water) if a spark is applied. This has the effect of supplying the activation energy for the reaction, which must always be considered in addition to arguments based on enthalpy changes.

Solution to worked example

This is the first problem in which we have been required to extract data from a table. The form of the headings used here may seem rather strange, but in fact this is the best way of presenting data. Let us look at the value of *b* for $O_{2(g)}$, the value in the very centre of the table. All we have to do is to equate the pure number in the table with its column heading, to give

$$\frac{b}{10^3 \text{ J K}^{-2} \text{ mol}^{-1}} = 12.98$$

Then

$$b = 12.98 \times 10^3 \text{ J K}^{-2} \text{ mol}^{-1}$$

and, if required, we can convert to correct scientific notation by rewriting as a number between 1 and 10 and adjusting the power of 10 as

$$b = 1.298 \times 10^4 \text{ J K}^{-2} \text{ mol}^{-1}$$

A similar technique is used to read values from graphs, whose axes labels should be given in a similar fashion to these table headings.

The answer to this problem will consist of another polynomial expression whose terms represent the difference in heat capacities between the reactants and the products. It will be of the form

$$\Delta C_p = \Delta a + T \Delta b + T^2 \Delta c$$

where

$$\Delta C_p = \sum C_p \text{ (products)} - \sum C_p \text{ (reactants)}$$

Notice that we are using notation (Δ and $\sum$) here to make the working more concise. Remember that Δ is used to represent a change, and that this is always taken as the final value minus the initial value. In this case, that is equivalent to taking products (final state) minus reactants (initial state). You may recall from section 1.3.3 that the $\sum$ symbol means 'add up', so we are simply summing the contributions from the products and from the reactants for each of the coefficients *a*, *b* and *c*. Therefore, we have

$$\Delta a = \sum a \text{ (products)} - \sum a \text{ (reactants)}$$
$$= 30.36 - [29.07 + (0.5 \times 25.72)] \text{ J K}^{-1} \text{ mol}^{-1}$$
$$= -11.57 \text{ J K}^{-1} \text{ mol}^{-1}$$

Note that the contribution from $O_{2(g)}$ is multiplied by 0.5 since this species appears in the reaction equation with a stoichiometric coefficient of $\frac{1}{2}$. Also

$$\Delta b = \sum b \text{ (products)} - \sum b \text{ (reactants)}$$

$$= 9.61 - [-0.836 + (0.5 \times 12.98)] \times 10^3 \text{ J K}^{-2} \text{ mol}^{-1}$$

$$\approx 3.96 \times 10^3 \text{ J K}^{-2} \text{ mol}^{-1}$$

and

$$\Delta c = \sum c \text{ (products)} - \sum c \text{ (reactants)}$$

$$= 11.8 - [20.1 + (0.5 \times -38.6)] \times 10^7 \text{ J K}^{-3} \text{ mol}^{-1}$$

$$= 11.0 \times 10^7 \text{ J K}^{-3} \text{ mol}^{-1}$$

It is slightly neater to write this using correct scientific notation as

$$\Delta c = 1.10 \times 10^8 \text{ J K}^{-3} \text{ mol}^{-1}$$

Now substituting into our equation for ΔC_p gives

$$\Delta C_p = -11.57 \text{ J K}^{-1} \text{ mol}^{-1} + (3.96 \times 10^3 \text{ J K}^{-2} \text{ mol}^{-1})T$$

$$+ (1.10 \times 10^8 \text{ J K}^{-3} \text{ mol}^{-1})T^2$$

Note the use of brackets to simplify this expression when the units are included. Another way of doing this would be to divide through by the common unit and write

$$\frac{\Delta C_p}{\text{J K}^{-1} \text{ mol}^{-1}} = -11.57 + (3.96 \times 10^3 \text{ K}^{-1})T + (1.10 \times 10^8 \text{ K}^{-2})T^2$$

Personal preference dictates how you should leave this expression.

2.2.3   Functions

So far in this section, we have seen that the heat capacity can be regarded as a constant in some circumstances, while in others we can specify a dependence on the temperature. In the latter case, we would say that the heat capacity is a function of temperature. It can then be denoted as $C_p(T)$, where the T in brackets indicates that the value of T determines that of C_p. In general, we use the notation $f(x)$ to denote a function f of the variable x.

As an example, if $f(x)$ is defined as

$$f(x) = 8x^2 + 2x + 9$$

we can calculate the value of the function for any value of x. Using this notation, $f(3)$ would be the value of this function when $x = 3$, and would be given by substituting for x on both sides of the defining equation to give

$$f(3) = (8 \times 3^2) + (2 \times 3) + 9$$

$$= (8 \times 9) + (2 \times 3) + 9$$

$$= 72 + 6 + 9$$

$$= 87$$

Worked example 2.4

If the heat capacity C_p of solid lead is given by the expression

$$C_p(T) = 22.13 \text{ J K}^{-1} \text{ mol}^{-1} + (1.172 \times 10^{-2} \text{ J K}^{-2} \text{ mol}^{-1})T$$
$$+ \frac{(9.6 \times 10^4 \text{ J K mol}^{-1})}{T^2}$$

determine the heat capacity at 298 K.

Chemical background

Observation showed that different quantities of heat were required to raise the temperature of equal amounts of different substances to the same extent. More careful experiments then showed that this heat capacity actually varied with the temperature. The SI unit of molar heat capacity is $\text{J K}^{-1} \text{ mol}^{-1}$; heat changes can be measured accurately by electrical means and $1 \text{ J} = 1 \text{ V} \times 1 \text{ C}$.

Solution to worked example

We actually need to calculate $C_p(298 \text{ K})$, which is given by a straightforward substitution into both sides of the defining equation.

$$C_p(298 \text{ K}) = 22.13 \text{ J K}^{-1} \text{ mol}^{-1} + (1.172 \times 10^{-2} \text{ J K}^{-2} \text{ mol}^{-1} \times 298 \text{ K})$$
$$+ \frac{9.6 \times 10^4 \text{ J K mol}^{-1}}{(298 \text{ K})^2}$$
$$\approx 22.13 \text{ J K}^{-1} \text{ mol}^{-1} + 3.492 \text{ J K}^{-1} \text{ mol}^{-1} + 1.081 \text{ J K}^{-1} \text{ mol}^{-1}$$
$$= (22.13 + 3.492 + 1.081) \text{ J K}^{-1} \text{ mol}^{-1}$$
$$\approx 26.70 \text{ J K}^{-1} \text{ mol}^{-1}$$

2.3
Clapeyron equation   There is a certain amount of vapour above every solid and liquid and if this is not free to escape, as in a closed vessel, equilibrium will be reached so that as

many molecules leave the surface as re-enter. At this equilibrium, the pressure above the solid or liquid is known as the **vapour pressure**. A study of vapour pressure is important because it can provide us with information about the solid or liquid.

The Clapeyron equation applies to any phase transition of a pure substance, and is expressed by the equation

$$\frac{\mathrm{d}p}{\mathrm{d}T} = \frac{\Delta H_t}{T\,\Delta V_t}$$

where ΔH_t and ΔV_t are the enthalpy and volume changes respectively which accompany a phase transition t at absolute temperature T. The left-hand side introduces a notation which we have not met before; it needs to be treated as a complete quantity and the letter d, on its own, has no meaning in this equation. The quantity

The graph of pressure against temperature for a substance is known as its **phase diagram**, and can be used to indicate whether a solid, liquid or gas will be found for given values of these variables. A typical one-component phase diagram is shown in Figure 2.3.

$$\frac{\mathrm{d}p}{\mathrm{d}T}$$

is called the **derivative of p with respect to T**. As we will see below, this function is equal to the gradient of the tangent of the graph of p against T.

2.3.1   Differentiation

If we consider a function such as

$$f(x) = 3x^2 + 1$$

whose graph is shown in Figure 2.4, we can calculate an **average rate of change** between any two values of x, such as $x = 1$ and $x = 5$. This is found by dividing the difference between the values of the function at the two values of x

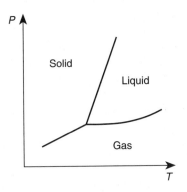

Figure 2.3   A typical phase diagram.

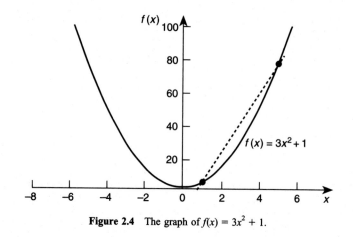

Figure 2.4   The graph of $f(x) = 3x^2 + 1$.

by the difference in the two values of x:

$$\text{average rate of change} = \frac{f(5) - f(1)}{5 - 1}$$

Since

$$f(1) = (2 \times 1^2) + 1 = 3 + 1 = 4$$

and

$$f(5) = (3 \times 5^2) + 1 = 75 + 1 = 76$$

we have

$$\text{average rate of change} = \frac{76 - 4}{5 - 1}$$

$$= \frac{72}{4}$$

$$= 18$$

This is equivalent to calculating the gradient of the line drawn between $f(1)$ and $f(5)$ on the graph shown in Figure 2.4.

It is also possible to determine the **instantaneous rate of change** of this function at a specified value of x, say $x = 2$. This is equivalent to determining the gradient of the tangent drawn to the curve at this value, as shown in Figure 2.5.

More generally, suppose that we wish to find the instantaneous rate of change, or gradient, when x has the specified value x_0. This is specified as point P in Figure 2.6. A second point Q on the curve defined by $f(x)$ has $x = x_0 + h$. The gradient of the line drawn between P and Q will therefore be

$$\frac{f(x_0 + h) - f(x_0)}{(x_0 + h) - x_0} = \frac{f(x_0 + h) - f(x_0)}{h}$$

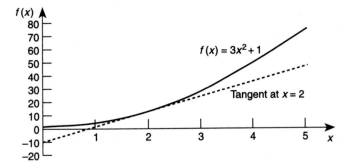

Figure 2.5   The gradient of the tangent to the function $f(x) = 3x^2 + 1$, at the point where $x = 2$.

Imagine now that Q slides down the curve until it is almost at the same place as P. This has the effect of making the value of h very small. In our minds, we can make h as small as we like as long as it does not become exactly zero. This is because if h were to become zero, the denominator would be zero, but division by zero is not defined. However, making h as close to zero as possible will give a very close approximation to the value of the gradient which we require. In mathematical terminology, we would say that we wish to calculate the above expression in the limit as h tends to zero. This is written as

$$\text{gradient of tangent} = \lim_{h \to 0} \frac{f(x_0 + h) - f(x_0)}{h}$$

More usually, we would write this as

$$\frac{\mathrm{d}f(x)}{\mathrm{d}x} = \lim_{h \to 0} \frac{f(x_0 + h) - f(x_0)}{h}$$

where

$$\frac{\mathrm{d}f(x)}{\mathrm{d}x}$$

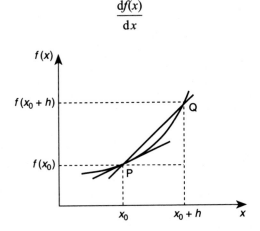

Figure 2.6   Definition of the derivative $\mathrm{d}f(x)/\mathrm{d}x$.

is the **derivative of $f(x)$ with respect to** x. This is spoken as 'd f of x by d x'. It is important to remember that this represents the gradient of a graph of $f(x)$ against x.

So far, the method of calculating a derivative involves taking a limit of a mathematical expression. In practice, this is not necessary as we can use a number of standard derivatives. In the case of a power of x, the derivative of x^n is nx^{n-1}, where n is a constant.

This table shows this rule for $n = 0, 1, 2, 3, 4$.

n	power of x	derivative
0	$x^0 = 1$	$0 \times x^{-1} = 0$
1	x^1	$1 \times x^0 = 1$
2	x^2	$2 \times x^1 = 2x$
3	x^3	$3 \times x^2 = 3x^2$
4	x^4	$4 \times x^3 = 4x^3$

When a power of x has a coefficient, e.g. ax^n, the derivative is anx^{n-1}, so that, for example,

$$\frac{d(3x^4)}{dx} = 3(4x^{4-1}) = 12x^3$$

Notice that a constant on its own can be thought of as a term where the constant is multiplied by an x^0. The derivative of x^0 will always be zero, so that, for instance

$$\frac{d(6)}{dx} = \frac{d(6x^0)}{dx} = \frac{6d(x^0)}{dx} = 6 \times 0 = 0$$

The rule given is readily extended to more complicated functions. If, for example, we have a function $f(x)$ defined as

$$f(x) = 2x^3 + 4x^2 + 7x + 2$$

we can simply apply the standard derivatives to each of these in turn. Notice the notation used for describing each of these derivatives:

$$\frac{d(2x^3)}{dx} = 2\frac{d(x^3)}{dx} = 2 \times 3x^2 = 6x^2$$

$$\frac{d(4x^2)}{dx} = 4\frac{d(x^2)}{dx} = 4 \times 2x = 8x$$

$$\frac{d(7x)}{dx} = 7\frac{d(x)}{dx} = 7 \times 1 = 7$$

$$\frac{d(2)}{dx} = 0$$

Putting this together gives the derivative as

$$\frac{df(x)}{dx} = 6x^2 + 8x + 7$$

Worked example 2.5

The triple point of iodine, I_2, occurs at a temperature of 114°C and a pressure of 12 kPa. If the enthalpies of fusion $\Delta_{fus}H$ and vaporization $\Delta_{vap}H$ are 15.52 kJ mol^{-1} and 41.80 kJ mol^{-1} respectively, sketch the pressure–temperature graph in the region of the triple point. Take the densities of the solid and liquid to be 4.930 g cm^{-3} and 2.153 g cm^{-3} respectively and the molar mass of iodine atoms to be 126.9 g mol^{-1}. Assume that iodine vapour behaves as an ideal gas, and that the enthalpy of sublimation $\Delta_{sub}H$ is given by the equation

$$\Delta_{sub}H = \Delta_{fus}H + \Delta_{vap}H$$

Chemical background

The term **sublimation** refers to a change directly from solid to vapour without passing through the liquid phase. Such behaviour can be predicted from the graph of pressure against temperature showing the equilibria between phases. From the equation given above, we can see that the enthalpy change will be the same when we pass from the solid to vapour phases, regardless of whether an intermediate liquid phase is involved.

The phase diagram obtained in this question has the three lines representing phase boundaries all with positive gradients. If we repeated the exercise for water, we would find that the slope of the line representing the solid to liquid transition would be negative. This is quite unusual, and is a result of the reduction in volume of ice on melting.

Solid carbon dioxide also sublimes

Solution to worked example

The question gives us all the information we need to evaluate the right-hand side of the Clapeyron equation

$$\frac{dp}{dT} = \frac{\Delta H_t}{T \Delta V_t}$$

which is equal to the left-hand side, i.e. the gradient of the graph of p and T. Since we are only asked about values close to the triple point, we can assume that the gradients we obtain are constant in this region of interest. We need to consider the three possible phase transitions in turn: solid to liquid; liquid to vapour; and solid to vapour.

Solid to liquid

The volume change ΔV will be

$$\Delta V = V_{\text{liquid}} - V_{\text{solid}}$$

where each volume can be obtained from the formula

$$V = \frac{m}{\rho}$$

with m representing mass and ρ density. Since 1 mol of I_2 has a mass of 2×126.9 g or 253.8 g, we have

$$V_{\text{liquid}} = \frac{253.8 \text{ g mol}^{-1}}{2.153 \text{ g cm}^{-3}} \simeq 117.9 \text{ cm}^3 \text{ mol}^{-1}$$

$$V_{\text{solid}} = \frac{253.8 \text{ g mol}^{-1}}{4.930 \text{ g cm}^{-3}} \simeq 51.48 \text{ cm}^3 \text{ mol}^{-1}$$

so that

$$\Delta V = (117.9 - 51.48) \text{ cm}^3 \text{ mol}^{-1} \simeq 66.4 \text{ cm}^3 \text{ mol}^{-1}$$

We also need to convert the given temperature value of 114°C to units of K.

$$\frac{T}{K} = \frac{114°C}{°C} + 273 = 387$$

so

$$T = 387 \text{ K}$$

The Kelvin temperature scale does not use negative numbers as its zero is the lowest possible temperature which can be attained. Temperatures measured on this scale are known as absolute temperatures (symbol T) and are related to those on the Celsius scale (symbol θ) by the expression $T/K = \theta/°C + 273.15$.

Note that the more precise conversion factor (273.15 rather than 273) is not used because the temperature is only given to the nearest °C.

Substituting into the Clapeyron equation now gives

$$\frac{dp}{dT} = \frac{15.52 \text{ kJ mol}^{-1}}{387 \text{ K} \times 66.4 \text{ cm}^3 \text{ mol}^{-1}}$$

$$= \frac{15.52 \times 10^3 \text{ J mol}^{-1}}{387 \text{ K} \times 66.4 \times 10^{-6} \text{ m}^3 \text{ mol}^{-1}}$$

$$\simeq 6.04 \times 10^5 \text{ Pa K}^{-1}$$

since $1 \text{ kJ} = 10^3 \text{ J}$, $1 \text{ cm}^3 = 10^{-6} \text{ m}^3$, and $1 \text{ Pa} = 1 \text{ N m}^{-2} = 1 \text{ Nm m}^{-3} = 1 \text{ J m}^{-3}$.

Note the conversion from centimetres to metres:

$$100 \text{ cm} = 1 \text{ m}$$

$$1 \text{ cm} = 10^{-2} \text{ m}$$

cubing each side gives us

$$(1 \text{ cm})^3 = (10^{-2} \text{ m})^3$$

or

$$1^3 \text{ cm}^3 = 10^{-6} \text{ m}^3$$

Notice that $(10^{-2})^3 = 10^{-6}$. Now $1^3 = 1$, so

$$1 \text{ cm}^3 = 10^{-6} \text{ m}^3$$

Liquid to vapour

Since the question tells us to treat the iodine vapour as an ideal gas, we can use the ideal gas equation to calculate its volume at the triple point. If

$$pV = nRT$$

this can be rearranged (by dividing both sides by p) to give

$$V = \frac{nRT}{p}$$

Substituting values for 1 mol at the triple point gives us

$$V = \frac{1 \text{ mol} \times 8.314 \text{ J K}^{-1} \text{ mol}^{-1} \times 387 \text{ K}}{12 \text{ kPa}}$$

$$= \frac{8.314 \times 387}{12} \frac{\text{J}}{\text{kPa}}$$

$$= 268.1265 \times \frac{\text{J}}{10^3 \text{ J m}^{-3}}$$

$$\simeq 0.2681 \text{ m}^3$$

taking the value of the ideal gas constant R as $8.314 \text{ J K}^{-1} \text{ mol}^{-1}$, and since $1 \text{ kPa} = 10^3 \text{ Pa} = 10^3 \text{ J m}^{-3}$. We are now able to calculate the volume change for the vaporization process for 1 mol:

$$\Delta V = V_{\text{vapour}} - V_{\text{liquid}}$$

$$= 0.2681 \text{ m}^3 \text{ mol}^{-1} - 117.9 \text{ cm}^3 \text{ mol}^{-1}$$

We need to convert these two values to the same units. Generally, it is preferable to work in base units (such as metres) rather than multiples (such as centimetres), so we choose to convert cm^3 to m^3, and use the rule $1 \text{ cm}^3 = 10^{-6} \text{ m}^3$. Then

$$117.9 \text{ cm}^3 = 117.9 \times 10^{-6} \text{ m}^3$$

$$= 1.179 \times 10^{-4} \text{ m}^3$$

If we compare this with the vapour volume of 0.2681 m^3, we see that it is negligible and we are quite justified in approximating the volume change on vaporization to be equal to the volume of the vapour. In this case, we will take ΔV as $0.2681 \text{ m}^3 \text{ mol}^{-1}$. Substituting into the Clapeyron equation leads us to

$$\frac{dp}{dT} = \frac{41.80 \text{ kJ mol}^{-1}}{387 \text{ K} \times 0.2681 \text{ m}^3 \text{ mol}^{-1}}$$

$$= \frac{41.80 \times 10^3 \text{ J mol}^{-1}}{387 \times 0.2681 \text{ K m}^3 \text{ mol}^{-1}}$$

$$\approx 402.9 \text{ J K}^{-1} \text{ m}^{-3}$$

$$= 402.9 \text{ J m}^{-3} \text{ K}^{-1}$$

$$= 402.9 \text{ N m m}^{-3} \text{ K}^{-1}$$

$$= 402.9 \text{ N m}^{-2} \text{ K}^{-1}$$

$$= 402.9 \text{ Pa K}^{-1}$$

Solid to vapour

Since the volume of the solid will be even less than that of the liquid, we are again justified in taking the volume change as being equal to the volume of the vapour, so we will set ΔV to $0.2681 \text{ m}^3 \text{ mol}^{-1}$. The enthalpy change $\Delta_{sub} H$ is given by the expression in the question, which evaluates to give

$$\Delta_{sub} H = \Delta_{fus} H + \Delta_{vap} H$$

$$= 15.52 \text{ kJ mol}^{-1} + 41.80 \text{ kJ mol}^{-1}$$

$$= 57.32 \text{ kJ mol}^{-1}$$

Substitution into the Clapeyron equation now gives

$$\frac{dp}{dT} = \frac{57.32 \text{ kJ mol}^{-1}}{387 \text{ K} \times 0.2681 \text{ m}^3 \text{ mol}^{-1}}$$

$$= \frac{57.32 \times 10^3 \text{ J mol}^{-1}}{387 \times 0.2681 \text{ K m}^3 \text{ mol}^{-1}}$$

$$\approx 552.5 \text{ Pa K}^{-1}$$

Using these three gradient values, we now need to draw lines of appropriate slope which pass through the triple point. Notice that the value of dp/dT for the fusion process is much larger than the other two, so this can be represented by a vertical line. The value for vaporization is

$$\frac{402.9}{552.5} = 0.729 \approx 0.7$$

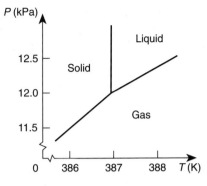

Figure 2.7   Phase diagram for iodine.

or 0.7 times the value for sublimation, which allows a simple qualitative representation to be drawn.

One possible graph, with values included, is shown in Figure 2.7.

2.4 Clausius–Clapeyron equation

The Clausius–Clapeyron equation relates the vapour pressures p_1 and p_2 at two temperatures T_1 and T_2 to the enthalpy of vaporization $\Delta_{vap}H$:

$$\ln\left(\frac{p_2}{p_1}\right) = \frac{\Delta_{vap}H}{R}\left[\left(\frac{1}{T_1}\right) - \left(\frac{1}{T_2}\right)\right]$$

Here we meet the ln symbol for the first time. This is an abbreviation for natural logarithm, and in this equation is telling us to 'take the natural logarithm of' the quantity p_2/p_1.

2.4.1   Logarithms

Historically, logarithms were used to make the processes of multiplication and division easier, but with the advent of electronic calculators this use is now redundant. However, their use is of much more fundamental importance than that, particularly when dealing with quantities which may span a very large range of values. There are two types of logarithms with which we need to be concerned: natural logarithms (ln), and logarithms to base 10 (log).

If three numbers a, b and c are related such that

$$a = b^c$$

we can write

$$\log_b a = c$$

where b is called the **base** of the logarithm. Thus, if we choose b to have the value 10, then if

$$a = 10^c$$

it follows that

$$\log a = c$$

since we normally use the term 'log' to denote logarithms to the base 10. On the other hand, if b is set to the number known as e, which has the value 2.718 28 ..., we have

$$a = e^c$$

and

$$\ln a = c$$

We will meet an example of the natural logarithm emerging from the analysis of a physical chemistry problem in Chapter 4. This frequently arises from expressions involving a term in $1/x$ where x is a variable.

since logarithms to the base 'e' are called 'natural logarithms' and are denoted by the term 'ln'.

Natural logarithms frequently arise from the mathematical analysis of problems in physical chemistry, while logarithms to the base 10 are of more use when we want to display a large range of data. The two types of logarithms are related by the expression

$$\ln x = 2.303 \log x$$

Worked example 2.6

The vapour pressure p of neon as a function of temperaturature θ is as follows:

θ	p
°C	mm Hg
−228.7	19 800
−233.6	10 040
−240.2	3170
−243.7	1435
−245.7	816
−247.3	486
−248.5	325

By using an appropriate mathematical transformation, display this data graphically.

Chemical background

These data are a selection of those which were obtained when this study was performed. The complete set was fitted to an expression of the form

$$\log\left(\frac{p}{\text{mm Hg}}\right) = \frac{0.05223A}{T} + B + CT$$

to give the values of the constants

$$A = -1615.5 \text{ K}$$

$$B = 5.699\,91$$

$$C = 0.011\,1800 \text{ K}^{-1}$$

Notice the use of the heading $\log(p/\text{mm Hg})$ in the equation above. This is because it is only possible to take the logarithm of a pure number, i.e. one which does not have any units. If all the pressure values are divided by the units, we end up with pure numbers. For example, if $p = 19\,800$ mm Hg, we can write

$$\frac{p}{\text{mm Hg}} = \frac{19\,800 \text{ mm Hg}}{\text{mm Hg}} = 19\,800$$

which is a similar technique to the one we used in Worked Example 2.3.

Solution to worked example

We will consider the details of graph drawing shortly, but for the moment simply note that pressure is plotted on the y-axis and temperature on the x-axis. This gives the graph shown in Figure 2.8, where we see that as the temperature increases the points become increasingly further apart and so the plot has very uneven spaces between the points. This leads to significant differences in the extent of precision to which the position of each point can be defined. Suppose that instead of plotting pressure, we take its logarithm to base 10 and plot that

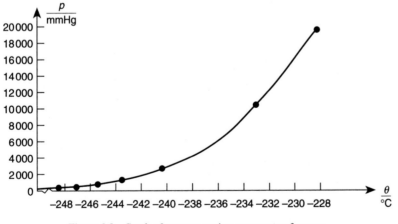

Figure 2.8   Graph of pressure against temperature for neon.

19800

against temperature. Using a calculator 🖩 gives us the following values:

$\dfrac{\theta}{°C}$	$\log\left(\dfrac{p}{\text{mm Hg}}\right)$
−228.7	4.297
−233.6	4.002
−240.2	3.501
−243.7	3.157
−245.7	2.912
−247.3	2.687
−248.5	2.512

Notice that the

$$\log\left(\frac{p}{\text{mm Hg}}\right)$$

values now occupy a much smaller range. From Figure 2.9, we can also see that they are far more evenly spread, and so easier to plot.

2.4.2 The equation of a straight line

We saw in section 2.2.2 that a polynomial of order 1 was known as a linear expression. Some examples are

$$5x + 2$$

$$3x + 7$$

$$2x + 2$$

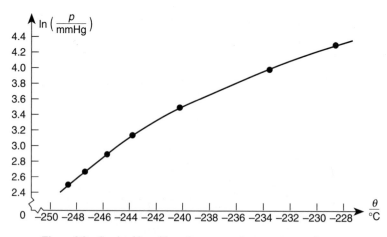

Figure 2.9 Graph of logarithm of pressure aginst temperature for neon.

These can be summarized by the general equation

$$y = mx + c$$

where y is simply the value we evaluate on the right-hand side of the equals sign, and m and c are constants. So if we write the first example as

$$y = 5x + 2$$

the constant m will be 5, and c will be 2. If we were to plot a graph of y against x, we would find that it was a straight line with gradient 5 and intercept 2, as shown in Figure 2.10. The consequence of this is that if we can rearrange any thermodynamic relationship into the form of a linear equation, we can obtain a straight line graph. Frequently, the values of the gradient and intercept of such graphs give us useful information about the system being studied.

Worked example 2.7

At the standard pressure, $p^{\ominus} = 1$ atm, a liquid boils at its normal boiling point, T_b. Given a set of values of the vapour pressure p at absolute temperature T, show how the Clausius–Clapeyron equation can be used to obtain the enthalpy of vaporization of the liquid.

Chemical background

A liquid boils when its vapour pressure is equal to the atmospheric or external pressure. More molecules are then leaving the surface of the liquid than are re-entering. Consequently, as the external pressure is reduced, the boiling temperature of the liquid falls.

The unit of pressure atmospheres (symbol atm) is defined as 101.325 kPa, and is a very practical unit to work with since 1 atm is atmospheric pressure. Typical normal boiling points are 100°C for water and 80.1°C for benzene, the former value being higher due to the presence of hydrogen bonds.

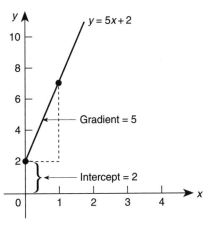

Figure 2.10   Graph of $y = 5x + 2$.

Solution to worked example

Starting from the Clausius–Clapeyron equation

$$\ln\left(\frac{p_2}{p_1}\right) = \frac{\Delta_{vap}H}{R}\left[\left(\frac{1}{T_1}\right) - \left(\frac{1}{T_2}\right)\right]$$

we can substitute the pair of values given into the equation. It is simpler to set $p_1 = p^{\ominus}$ and $T_1 = T_b$, and then to replace the symbols p_2 and T_2 by p and T respectively. Doing this gives the expression

$$\ln\left(\frac{p}{p^{\ominus}}\right) = \frac{\Delta_{vap}H}{R}\left[\left(\frac{1}{T_b}\right) - \left(\frac{1}{T}\right)\right]$$

At this stage, it is important to realize that p and T are the values of our variable pressure and temperature respectively, while $\Delta_{vap}H$, R, $p^{\ominus}$ and T_b are all constants. If we expand the right-hand side of the equation (by removing brackets) we have

$$\ln\left(\frac{p}{p^{\ominus}}\right) = \left(\frac{\Delta_{vap}H}{RT_b}\right) - \left(\frac{\Delta_{vap}H}{RT}\right)$$

which can be written as

$$\ln\left(\frac{p}{p^{\ominus}}\right) = -\frac{\Delta_{vap}H}{R}\left(\frac{1}{T}\right) + \left(\frac{\Delta_{vap}H}{RT_b}\right)$$

Consider what happens if we make the substitutions

$$x = \frac{1}{T} \qquad y = \ln\left(\frac{p}{p^{\ominus}}\right)$$

$$m = -\frac{\Delta_{vap}H}{R} \qquad c = \frac{\Delta_{vap}H}{RT_b}$$

You should find that this gives back the general equation

$$y = mx + c$$

If we plot $\ln(p/p^{\ominus})$ on the y-axis and $1/T$ on the x-axis, a straight line will be obtained with gradient $-\Delta_{vap}H/R$ and intercept $\Delta_{vap}H/RT_b$.

Since the value of the ideal gas constant R is known, the gradient gives the value of $\Delta_{vap}H$. Notice that we need to plot $\ln(p/p^{\ominus})$ on the y-axis. Since $p^{\ominus} = 1$ atm, if the values of p are also expressed in atm we will automatically be taking the natural logarithm of pure numbers, as required.

2.4.3   Plotting graphs

Having seen how to determine the plot we require, it is worth spending some time considering the mechanics of graph plotting. While there are now many computer programs available for doing this, they can give unusual results and have difficulties with certain data sets. It is also worth discussing manual graph plotting because the rules which govern overall appearance should also be applied to computer generated plots.

If you have an equation relating two variables, you should be able to determine which variable to plot on the x- and y-axes. However, there will be times when you will be instructed to 'plot variable 1 against variable 2'. This is *always* equivalent to plotting 'y against x', and it is often important to display them the right way around.

Having established this, the most important thing to remember is to fill as much of the graph paper as possible. It is best to have one plot per page, except when superimposing plots using the same axes for both, which can sometimes be useful. Try to avoid scales involving numbers such as 3 and 7 since these make it difficult to work out exactly where intermediate values should be plotted. Use a sharp pencil, label the axes and, if appropriate, include a title for the graph. There is no need to include the origin on either axis unless the spread of data lends itself to doing so. However, it is important not to mislead the reader so the following suggestion should be considered: if one axis includes zero, the other axis should intersect at the zero point, and any stretch of axis to the first label should be shown as a zigzag as shown below.

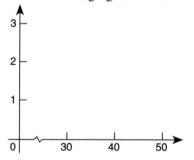

Worked example 2.8

The vapour pressure p of water is given below as a function of temperature θ. Determine the enthalpy of vaporization of water.

$\dfrac{\theta}{°C}$	$\dfrac{p}{10^{-2}\,\text{atm}}$
20	2.31
30	4.19
40	7.28
50	12.2
60	19.7
70	30.8

Chemical background

Typical values of the standard
enthalpy of vaporization are
8.2 kJ mol^{-1} for methane and
43.5 kJ mol^{-1} for ethanol; the
latter is higher due to the
presence of hydrogen bonding in
the liquid

The vapour pressure increases with temperature as a result of intermolecular forces in the liquid being broken. Vapour pressure measurements are consequently able to give us information about the nature of those forces. Most liquids with high values of enthalpy of vaporization also have high normal boiling points, although this is not always the case.

Solution to worked example

We saw in the previous problem that we need to plot $\ln(p/p^{\ominus})$ against $1/T$ where $p^{\ominus} = 1$ atm. It is important to realize that T refers to the absolute temperature (units of K) rather than the values given here. The first stage is to draw up a table containing the transformed data. If we consider the first pair of values, we have

$$\theta = 20°C$$

so that

$$T = (20 + 273) = 293 \text{ K}$$

 and

$$\frac{1}{T} = \frac{1}{293 \text{ K}}$$

$$\simeq 3.41 \times 10^{-3} \text{ K}^{-1}$$

$$p = 2.31 \times 10^{-2} \text{ atm}$$

so that

$$\frac{p}{p^{\ominus}} = \frac{2.31 \times 10^{-2} \text{ atm}}{1 \text{ atm}}$$

$$= 2.31 \times 10^{-2}$$

and

$$\ln\left(\frac{p}{p^{\ominus}}\right) = \ln(2.31 \times 10^{-2})$$

$$\simeq -3.77$$

Applying a similar treatment to the other values leads to the final table:

$\dfrac{T^{-1}}{10^{-3}\,\text{K}^{-1}}$	$\ln\!\left(\dfrac{p}{p^{\ominus}}\right)$
3.41	−3.77
3.30	−3.17
3.19	−2.62
3.10	−2.10
3.00	−1.63
2.92	−1.18

A suitable range of values on the x-axis (T^{-1}) would be from 2.9 to 3.5, with divisions of 0.1. On the y-axis ($\ln(p/p^{\ominus})$) it would be from −3.5 to −1.0, with divisions of 0.5. Exactly how these are fitted onto the graph paper depends on the particular arrangement of ruled squares. The labels on the axes are exactly as in the table above, to ensure that the points on the graph actually represent pure numbers. To obtain the gradient, we need to measure the increase in y and divide it by the corresponding increase in x, making sure that we use as large a portion of the straight line as possible. Using the values shown on the graph in Figure 2.11, this gives us

$$\text{gradient} = \frac{-1.18 - (-3.77)}{(2.92 - 3.41) \times 10^{-3}\,\text{K}^{-1}}$$

$$= \frac{2.59}{-4.9 \times 10^{-4}\,\text{K}^{-1}}$$

$$\simeq -5.29 \times 10^{3}\,\text{K}$$

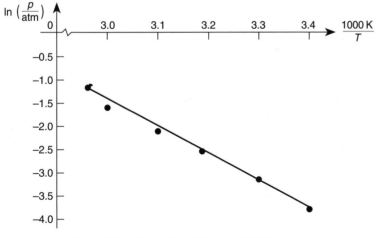

Figure 2.11   Graph of $\ln(p/p^{\ominus})$ against $1/T$ for water.

Notice that, in the first line of gradient calculation, the corresponding values of T^{-1} and $\ln(p/p^{\ominus})$ appear under one another. We know that the value of the gradient is equal to $-\Delta_{\mathrm{vap}}H/R$ so

$$-\frac{\Delta_{\mathrm{vap}}H}{R} = -5.29 \times 10^3 \text{ K}$$

Multiplying both sides of the equation by R and multiplying both sides by -1 to change the negative signs to plus signs on both sides gives

$$\Delta_{\mathrm{vap}}H = 5.29 \times 10^3 \text{ K} \times R$$

which on substituting $R = 8.314 \text{ J K}^{-1} \text{ mol}^{-1}$ gives

$$\Delta_{\mathrm{vap}}H = 5.29 \times 10^3 \text{ K} \times 8.314 \text{ J K}^{-1} \text{ mol}^{-1}$$

$$\simeq 44.0 \times 10^3 \text{ J mol}^{-1}$$

This can be written more concisely by using kJ rather than 10^3 J as

$$\Delta_{\mathrm{vap}}H = 44.0 \text{ kJ mol}^{-1}$$

2.5
The ideal gas equation

The ideal gas equation is the relationship you will probably meet most frequently in thermodynamics. This is because it describes the simplest form of matter to study, and is itself a simple expression. The volume of the molecules in an ideal gas would be negligible as would the forces between them. Real gases behave like ideal gases and obey the ideal gas equation at low pressures and high temperatures.

The ideal gas equation usually appears in the form

$$pV = nRT$$

where p is the pressure, V the volume, n the amount of gas, R the ideal gas constant and T the absolute temperature.

2.5.1   Interconversion of units

We have already met several examples in which units have been substituted along with the corresponding numerical value in an equation. This practice is essential, and allows us to obtain the correct units of the final calculated quantity in a systematic way. A similar technique may also be used to convert a quantity from one set of units to another.

We have already seen that for practical purposes we may wish to work with the pressure units of atm rather than kPa. Other non-SI units in common use are the angstrom (Å) which is equal to 10^{-10} m and the calorie (cal) which is equal to 4.184 J. The use of such units in certain fields is so well established that to try to change would probably be more confusing than learning to live with them.

Worked example 2.9

The ideal gas constant R is usually given as $8.314 \text{ J K}^{-1} \text{ mol}^{-1}$. What is its value when expressed in the units $\text{dm}^3 \text{ atm K}^{-1} \text{ mol}^{-1}$?

Chemical background

The value of the ideal gas constant was formerly determined from measurements of the molar volumes of oxygen and nitrogen, but more recently this has been done by determining the speed of sound in argon. Such changes in the way in which the values of constants are determined do take place from time to time, but invariably they lead to very small changes in their numerical values which are only significant in high precision work.

Solution to worked example

To solve this problem, we need to use the following conversion factors:

$$1 \text{ atm} = 101.325 \text{ kPa} = 101.325 \times 10^3 \text{ Pa}$$

$$1 \text{ J} = 1 \text{ N m}$$

$$1 \text{ Pa} = 1 \text{ N m}^{-2}$$

Comparing the units of the two forms of R shows that we essentially have to convert J to dm^3 atm. we do this by successive substitution in the original expression for R.

$$R = 8.314 \text{ J K}^{-1} \text{ mol}^{-1}$$

$$= 8.314 \text{ N m K}^{-1} \text{ mol}^{-1}$$

Notice that to obtain units of atm, we need to convert from Pa. At this stage, it helps to realize that

$$1 \text{ Pa} = 1 \text{ N m}^{-2} \quad \text{and} \quad 1 \text{ J} = 1 \text{ N m}^{-2} \text{ m}^3$$

so that

$$R = 8.314 \text{ N m}^{-2} \text{ m}^3 \text{ K}^{-1} \text{ mol}^{-1}$$

$$= 8.314 \text{ Pa m}^3 \text{ K}^{-1} \text{ mol}^{-1}$$

Rearranging the definition for 1 atm and substituting 10 dm = 1 m gives us

$$R = \frac{8.314 \text{ atm}}{101.325 \times 10^3} \times (10 \text{ dm})^3 \text{ K}^{-1} \text{ mol}^{-1}$$

$$\approx 8.205 \times 10^{-5} \text{ atm} \times 10^3 \text{ dm}^3 \text{ K}^{-1} \text{ mol}^{-1}$$

$$= 8.205 \times 10^{-2} \text{ dm}^3 \text{ atm K}^{-1} \text{ mol}^{-1}$$

2.5.2 Constants and variables

The ideal gas constant R is known as a **universal constant**, which means that its value does not change under any circumstances. This is in contrast to quantities such as n in the ideal gas equation. This is the amount of gas, which will be fixed for a given sample of gas in a closed container, but which may vary from sample to sample. The quantities p, V and T are variables, but any of them may be considered to be fixed according to experimental conditions.

Worked example 2.10

Identify the constants and variables in the following equations which also describe the behaviour of gases.

(a) The van der Waals equation is

$$\left(p + \frac{an^2}{V^2}\right)(V - nb) = nRT$$

Here p is the pressure of gas, V its volume, T the temperature and n the amount present. For carbon monoxide, $a = 0.1505$ Pa m^6 mol^{-2} and $b = 0.0398 \times 10^{-3}$ m^3 mol^{-1}.

(b) The Beattie–Bridgeman equation is

$$p = \frac{RT\left[1 - \left(\frac{c}{V_m T^3}\right)\right]}{V_m^2}(V_m + B) - \frac{A}{V_m^2}$$

with

$$A = A_0\left[1 - \left(\frac{a}{V_m}\right)\right]$$

and

$$B = B_0\left[1 - \left(\frac{b}{V_m}\right)\right]$$

Here p is the pressure of gas, V_m its molar volume and T the temperature. For methane, $A_0 = 0.230\,71$ Pa m^6 mol^{-1}, $a = 18.55 \times 10^{-6}$ m^3 mol^{-1}, $B_0 = 55.87 \times 10^{-6}$ m^3 mol^{-1}, $b = -15.87 \times 10^{-6}$ m^3 mol^{-1}, $c = 12.83 \times 10$ m^3 mol^{-1}.

(c) The virial equation is

$$pV_m = RT + Bp$$

Here p is the pressure of gas, V_m its molar volume, R the ideal gas constant and T the absolute temperature. For nitrogen, the second virial coefficient B is -4.2 cm^3 mol^{-1} at 300 K.

Chemical background

(a) The van der Waals equation was developed to account for the behaviour of a gas when it is liquefied. The volume term is reduced to allow for the finite volume of the molecules, while the pressure term is increased to allow for the presence of forces between molecules.
(b) The inclusion of a large number of parameters in the Beattie–Bridgeman equation allows its use in situations where a precise fit to experimental data is required, particularly at high pressures.
(c) The form of the virial equation given here actually comes from a power series of the form

$$pV_m = RT + Bp + Cp^2 + Dp^3 + \cdots$$

where B, C and D are functions of temperature and are known as the second, third and fourth virial coefficients respectively. One of the uses of this equation is that these coefficients may be related to the potential functions which describe the interaction between molecules. It is common to ignore terms in p^2 and higher.

Solution to worked example

(a) As in the case of the ideal gas equation, the variables are p, V and T, if the amount of gas n is fixed. The values of a and b are constant, depending only on the nature of gas being considered.
(b) This is also similar to the ideal gas equation in that p, V_m and T may vary. The other quantities A_0, a, B_0, b and c are constants and again depend only on the nature of the gas.
(c) As previously p, V_m and T are variables and R is a constant. However, the question suggests that B is a function of temperature so this could also be regarded as a variable, although its value at given temperatures may be obtained from tables.

2.5.3 Proportion

We can distinguish two types of proportion. If two quantities are **directly proportional** to one another, the ratio of their values will remain constant. For example, if x and y can have the following pairs of values they are in direct proportion:

x	1	2	3	4
y	2	4	6	8

In this case the ratio y/x remains constant at 2. We say that y is directly proportional to x. This is written mathematically as

$$y \propto x$$

where $\propto$ is the proportionality sign. If this is the case, we can introduce a constant of proportionality K and write the relationship as

$$y = Kx$$

Rearranging this equation gives

$$K = \frac{y}{x}$$

and we see that the constant of proportionality is simply the constant ratio of the two quantities x and y.

Conversely, x and y may be **inversely proportional** to one another. An example of such a relationship is in the following table:

x	1	2	3	4
y	24	12	8	6

This time, the product xy remains constant at 24. In this case, we would write

$$y \propto \frac{1}{x}$$

which on replacing $\propto$ by a new proportionality constant K' gives

$$y = \frac{K'}{x}$$

which rearranges to

$$K' = xy$$

so K' is the constant product of x and y.

Worked example 2.11

For a fixed amount of gas which obeys the ideal gas equation, describe the relationship between the two variables indicated when the third is held constant:

(a) p and V at constant T
(b) p and T at constant V
(c) T and V at constant p.

Chemical background

The experimental relationships between these variables are expressed by Boyle's Law and Charles's Law. Together with Avogadro's Law, these can be used to formulate the ideal gas equation. This problem involves deducing the original relationships from the final equation.

Solution to worked example

To solve this problem we need to be able to rearrange the ideal gas equation

$$pV = nRT$$

so that the two variables are on opposite sides of the equation. In each case, we will call the grouping of constant terms K, which will be our constant of proportionality.

(a) Dividing both sides of the equation by V gives

$$p = \frac{nRT}{V}$$

and since nRT is a constant, which we will call K, we can write

$$p = \frac{K}{V}$$

or

$$p \propto \frac{1}{V}$$

so p and V are inversely proportional to one another. This is Boyle's Law.

(b) Using the rearranged equation from part (a)

$$p = \frac{nRT}{V}$$

and since nR/V is a constant, which we will call K', we can write

$$p = K'T$$

or

$$p \propto T$$

so p and T are directly proportional.

(c) Dividing both sides of the original equation by p gives

$$V = \frac{nRT}{p}$$

and since nR/p is a constant, which we will call K'', we can write

$$V = K''T$$

or

$$V \propto T$$

so V and T are directly proportional. This is Charles's Law.

2.5.4 Functions of two variables

In section 2.2 we met the idea of writing an expression as a function of a variable, using the $f(x)$ notation. This can be extended to cases where we have two (or more) variables, such as

$$f(x, y) = x^3 + 3x^2y + 2y^2$$

and

$$g(x, y) = \ln(3x) + 2xy$$

Worked example 2.12

Use the function notation to write expressions for the

(a) pressure
(b) volume, and
(c) temperature

of an ideal gas.

Chemical background

Whereas we can display a function of one variable as a line graph, when two variables are involved, this is not possible. One solution is to use a **contour map**, where the height above the axes gives the value of the functions. This has been done for the pressure, volume and temperature of an ideal gas in Figure 2.12.

Solution to worked example

(a) We can use the rearranged equation from part (a) of Worked Example 2.11. Since both temperature and volume are variables, we write

$$p(T, V) = \frac{nRT}{V}$$

(b) We can use the rearranged equation from part (c) of Worked Example 2.11. Since both temperature and pressure are variables, we write

$$V(T, p) = \frac{nRT}{p}$$

(c) We can rearrange the ideal gas equation (by dividing both sides by nR) to make T the subject:

$$T = \frac{pV}{nR}$$

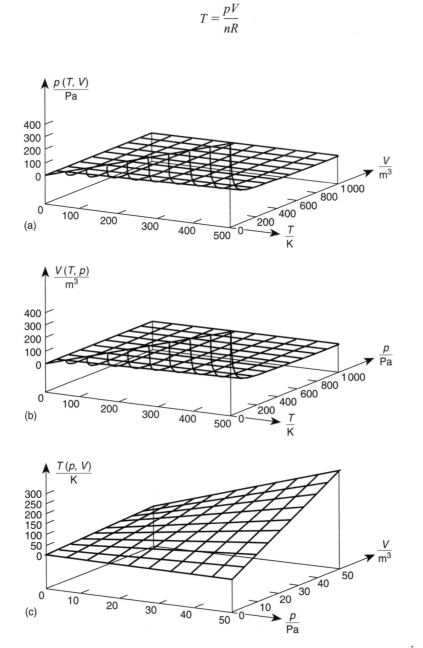

Figure 2.12 Contour maps representing the ideal gas equation. Units are Pa for pressure p, m^3 for volume V and K for absolute temperature T.

Since p and V are both variables

$$T(p, V) = \frac{pV}{nR}$$

2.5.5   Partial differentiation

We have already met the idea of differentiating a function of one variable in section 2.3. This was equivalent to determining the gradient of the curve representing that function, and represented the change in the function for a given change in the variable.

When we have a function of two variables, changing either of them will produce a change in the value of the function. This change is given by a quantity called the **partial derivative**. If our function is $f(x, y)$, then it is possible to determine

$$\left(\frac{\partial f(x, y)}{\partial x} \right)_y$$

the partial derivative with respect to x when y is held constant and

$$\left(\frac{\partial f(x, y)}{\partial y} \right)_x$$

the partial derivative with respect to y when x is held constant. Note that, in both cases, the quantity to be held constant is shown outside the bracket. The notation for each derivative is often abbreviated to

$$\frac{\partial f}{\partial x} \quad \text{and} \quad \frac{\partial f}{\partial y}$$

respectively for clarity and to save space.

The quantity $\partial f / \partial x$ represents the change in $f(x, y)$ when x is changed, and is equal to the gradient of the tangent to the surface of $f(x, y)$ in the x-direction, as shown in Figure 2.13 where the tangent is drawn at a constant value of y, denoted as k. Similarly, $\partial f / \partial y$ represents the change in $f(x, y)$ when y is changed, and is equal to the gradient of the tangent to the surface $f(x, y)$ in the y-direction.

Partial derivatives are calculated in the same way as full derivatives, while holding the fixed variable constant. If a function is defined so that

$$f(x, y) = 2x^2 + 3xy + 2xy^2$$

we can rewrite this treating every term in y as a constant. In the expression below, all the constant terms are enclosed in brackets:

$$f(x, y) = (2)x^2 + (3y)x + (2y^2)x$$

Differentiating x^2 gives $2x$ and differentiating x gives 1, so we obtain

$$\frac{\partial f}{\partial x} = (2)2x + (3y)1 + (2y^2)1$$

$$= 4x + 3y + 2y^2$$

Similarly, if we rewrite the expression for $f(x, y)$ treating every term in x as a constant then

$$f(x, y) = (2x^2) + (3x)y + (2x)y^2$$

Differentiating a constant gives 0, differentiating y gives 1 and differentiating y^2 gives $2y$. We therefore obtain

$$\frac{\partial f}{\partial y} = 0 + (3x)1 + (2x)2y$$

$$= 3x + 4xy$$

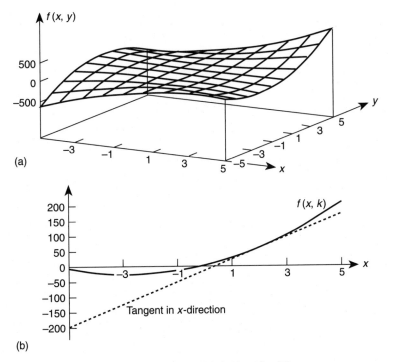

Figure 2.13   The partial derivative $\partial f(x, y)/\partial x$.

Worked example 2.13

For an ideal gas, calculate the partial derivatives:

$$\frac{\partial p}{\partial V} \quad \text{and} \quad \frac{\partial p}{\partial T}$$

Chemical background

The other Maxwell relations are:

$$\left(\frac{\partial T}{\partial V}\right)_{S} = -\left(\frac{\partial p}{\partial S}\right)_{V}$$

$$\left(\frac{\partial T}{\partial p}\right)_{S} = \left(\frac{\partial V}{\partial S}\right)_{p}$$

$$\left(\frac{\partial p}{\partial T}\right)_{V} = \left(\frac{\partial S}{\partial V}\right)_{T}$$

Partial derivatives find much use in thermodynamics, as it is possible to derive relationships between them. One example of this is Maxwell's relations, which consist of four relationships of the type

$$\left(\frac{\partial V}{\partial T}\right)_{p} = -\left(\frac{\partial S}{\partial p}\right)_{T}$$

These are useful because, while it is easy to measure the variation of volume V with temperature T, it is not straightforward to measure entropy S as a function of pressure p in the laboratory.

Solution to worked example

We can use the expression obtained in part (a) of Worked Example 2.12:

$$p(T, V) = \frac{nRT}{V}$$

To calculate the partial derivative with respect to V, we treat T as a constant, so the expression on the right-hand side can be written with the constant term in brackets as

$$p(T, V) = (nRT) \cdot \frac{1}{V}$$

$$= (nRT)V^{-1}$$

Remembering that we differentiate by multiplying by the power and reducing the power by 1, x^{-1} would differentiate (with respect to x) to give $(-1)x^{-2}$, i.e.

$$\frac{dx^{-1}}{dx} = (-1)x^{-1-1} = -\frac{1}{x^{2}}$$

Therefore we obtain

$$\frac{\partial p}{\partial V} = nRT\left(\frac{-1}{V^2}\right)$$

$$= -\frac{nRT}{V^2}$$

To calculate the partial derivative with respect to T, we treat V as a constant, so grouping the constant term in brackets gives us

$$p(T, V) = \left(\frac{nR}{V}\right)T$$

Remember that differentiating x with respect to x gives us 1, i.e.

$$\frac{dx}{dx} = 1$$

Here we are differentiating with respect to T, which gives

$$\frac{\partial p}{\partial T} = \left(\frac{nR}{V}\right) \times 1$$

$$= \frac{nR}{V}$$

2.5.6   The differential

We saw, in Worked Example 2.8, that we could determine the gradient m of a straight line graph of y against x as

$$m = \frac{\text{increase in } y}{\text{increase in } x} = \frac{\Delta y}{\Delta x}$$

It is possible to rearrange this equation to give an expression for the increase in either of the variables x or y:

$$\Delta x = \frac{\Delta y}{m} \qquad \Delta y = m\,\Delta x$$

Similarly, we saw in section 2.3 that the gradient of a curve is given by the derivative dy/dx which comes from considering very small or infinitesimal changes in the variables x and y. By analogy with our treatment for the straight line above, we can write

$$dy = m\,dx$$

where the gradient m will be a function of x. This expression allows us to express the change in variable y in terms of the variable x and the change in x. For example, if

$$y = 2x^2 + x + 3$$

the derivative

$$\frac{dy}{dx} = 4x + 1$$

and so

$$dy = (4x + 1)\, dx$$

where dy is known as the **differential**.

This approach can be readily extended to functions of more than one variable, such as $f(x, y)$. The differential $df(x, y)$ is then defined by

$$df(x, y) = \left(\frac{\partial f}{\partial x}\right)_y dx + \left(\frac{\partial f}{\partial y}\right)_x dy$$

This can be interpreted as giving the change in the function $f(x, y)$ when the variables x and y are changed by small amounts dx and dy respectively. For example, if

$$f(x, y) = xy^3 + 2x^2y$$

we calculate the partial derivative $\partial f/\partial x$ by treating y as a constant. Then

$$f(x, y) = (y^3)x + (2y)x^2$$

and

$$\frac{\partial f}{\partial x} = (y^3)1 + (2y)2x$$

$$= y^3 + 4xy$$

Treating x as a constant allows us to calculate the partial derivative $\partial f/\partial y$:

$$f(x, y) = (x)y^3 + (2x^2)y$$

and

$$\frac{\partial f}{\partial y} = (x)3y^2 + (2x^2)1$$

$$= 3xy^2 + 2x^2$$

Substituting in our general expression above for the differential gives:

$$df = (y^3 + 4xy)\, dx + (3xy^2 + 2x^2)\, dy$$

Worked example 2.14

Calculate the differential dp of an ideal gas.

Chemical background

Differentials can be exact or inexact. For an exact differential, the relationship

$$\left[\frac{\partial}{\partial y}\left(\frac{\partial f}{\partial x}\right)_y\right]_x = \left[\frac{\partial}{\partial x}\left(\frac{\partial f}{\partial y}\right)_x\right]_y$$

must hold. If this is the case, the function $f(x, y)$ is known as a **state function**. This means that its value is independent of the path taken to get to the particular state. This idea was discussed more fully in section 2.1.

Solution to worked example

We first need to write the pressure p of the ideal gas as a function of the other variables by rearranging the ideal gas equation. Since $pV = nRT$ we have

$$p = \frac{nRT}{V}$$

Since the variables on the right of this equation are the temperature T and the volume V, it may be clearer to write this using our notation for a function of two variables:

$$p(T, V) = \frac{nRT}{V}$$

To determine the differential dp, we need to calculate the partial derivatives

$$\left(\frac{\partial p}{\partial T}\right)_V \quad \text{and} \quad \left(\frac{\partial p}{\partial V}\right)_T$$

These were determined in Worked Example 2.13, and were

$$\left(\frac{\partial p}{\partial T}\right)_V = \frac{nR}{V} \quad \text{and} \quad \left(\frac{\partial p}{\partial V}\right)_T = -\frac{nRT}{V^2}$$

The differential dp will be given by the expression

$$dp = \left(\frac{\partial p}{\partial T}\right)_V dT + \left(\frac{\partial p}{\partial V}\right)_T dV$$

so we can substitute directly and obtain

$$dp = \left(\frac{nR}{V}\right) dT - \left(\frac{nRT}{V^2}\right) dV$$

2.6
The van der Waals equation

We have already met the van der Waals equation to describe the behaviour of a gas in Worked Example 2.10, where we saw that corrections were made to both the pressure and volume to improve the description of real gases. These are the constants a and b in the equation

$$\left(p + \frac{an^2}{V^2}\right)(V - nb) = nRT$$

in which all the other symbols have the same meaning as in the ideal gas equation.

2.6.1 Expansion of brackets

The maths involved in the expansion of expressions using brackets has been discussed in section 2.2.1.

Worked Example 2.15

Remove the brackets from the van der Waals equation.

Chemical background

The constants a and b in the van der Waals equation are normally obtained by fitting the equation to experimental measurements of pressure, volume and temperature.

Alternatively, since there is a critical temperature T_c above which a gas cannot be liquefied by pressure alone, it is possible to define a corresponding critical pressure p_c and critical volume V_c. The quantities are related to the van der Waals constants by the equations

$$a = 3p_c V_c^2 \quad \text{and} \quad b = \frac{V_c}{3}$$

Solution to worked example

Multiplying the second bracket by each of the terms in the first bracket, gives

$$p(V - nb) + \left(\frac{an^2}{V^2}\right)(V - nb) = nRT$$

Each term inside the remaining brackets of $(V - nb)$ can now be multiplied by the term outside, to give

$$pV - pnb + \left(\frac{an^2}{V^2}\right)V - \left(\frac{an^2}{V^2}\right)nb = nRT$$

We can now simplify the third and fourth terms on the left-hand side of this equation:

$$\left(\frac{an^2}{V^2}\right)V = \frac{an^2V}{V^2} = \frac{an^2}{V} \quad \text{and} \quad \left(\frac{an^2}{V^2}\right)nb = \frac{an^3b}{V^2}$$

so that the overall equation becomes

$$pV - pnb + \frac{an^2}{V} - \frac{an^3b}{V^2} = nRT$$

This would be a reasonable way of leaving the equation, as the brackets have been removed and all the terms simplified. However, a more elegant solution would be to continue as follows. The nRT term on the right can be 'removed' by subtracting nRT from both sides of the equation to give

$$pV - pnb + \frac{an^2}{V} - \frac{an^3b}{V^2} - nRT = 0$$

If we now multiply each term by V^2 (noting that $0 \times V^2 = 0$) we obtain

$$pV^3 - pnbV^2 + an^2V - an^3b - nRTV^2 = 0$$

or

$$pV^3 - (pnb + nRT)V^2 + an^2V - an^3b = 0$$

which can be classified as a cubic expression in V and contains no fractions.

2.6.2   Combining limits

The concept of a limit was discussed in section 2.3. Limits can be combined in much the same way as the quantities they represent. For example, if

$$f(x) = g(x) + h(x)$$

then

$$\operatorname*{Lim}_{x \to L} f(x) = \operatorname*{Lim}_{x \to L} g(x) + \operatorname*{Lim}_{x \to L} h(x)$$

Similarly, if

$$f(x) = g(x)\, h(x)$$

then

$$\operatorname*{Lim}_{x \to L} f(x) = \operatorname*{Lim}_{x \to L} g(x) \operatorname*{Lim}_{x \to L} h(x)$$

Worked example 2.16

Show that the van der Waals equation reduces to the ideal gas equation in the limit of low pressure.

Chemical background

A general requirement of all acceptable equations of state for gases is that they will reduce to the ideal gas equation as the pressure falls to zero. The interactions between molecules are then negligible, as is the volume of the molecules relative to the volume of the gas, so we are approaching the ideal situation.

Solution to worked example

In terms of limits, we can write the van der Waals equation as

$$\text{Lim}_{p \to 0} \left(p + \frac{an^2}{V^2} \right) \text{Lim}_{p \to 0} (V - nb) = \text{Lim}_{p \to 0} nRT$$

The first term to consider is

$$p + \frac{an^2}{V^2}$$

in the first bracket. We need to realize that, as the pressure p tends towards zero, the volume V tends towards infinity. This can be expressed mathematically as $p \to 0, V \to \infty$.

As V becomes very large, an^2/V^2 will tend towards zero. Also because the V term is squared, this will happen faster than p tends to zero. Thus this term will tend to p, or mathematically

$$\text{Lim}_{p \to 0} \left(p + \frac{an^2}{V^2} \right) = \text{Lim}_{V \to \infty} \left(p + \frac{an^2}{V^2} \right) = p$$

Now we consider the term in the second bracket. Again, as $p \to 0, V \to \infty$ so

$$\text{Lim}_{p \to \infty} (V - nb) = \text{Lim}_{V \to \infty} (V - nb)$$

As V becomes very large, the quantity nb remains the same, and so the value of the second limit becomes closer to V:

$$\text{Lim}_{p \to 0} (V - nb) = V$$

Finally, as neither p nor V appears on the right of the equation, we have

$$\text{Lim}_{p \to 0} (nRT) = nRT$$

Substituting into our equation involving limits, this now gives us

$$pV = nRT$$

which is the ideal gas equation, as required.

The equilibrium constant is a fundamental quantity in thermodynamics, describing the equilibrium state of a system in terms of concentration units. Its magnitude determines the extent to which a particular reaction will take place; a large value indicates a greater concentration of products than reactants and hence a favourable forward reaction.

A mathematical way of defining the equilibrium constant K is to use the equation

$$K = \prod a_j^{v_j}$$

The symbol a_j here is used to represent the activity of species j; activities are the correct way of defining dimensionless equilibrium constants, i.e. those without units. The mathematical symbol $\prod$ is new to us. It simply means 'multiply', or 'take the product'. The symbol v_j is known as the stoichiometric number of species j. This is obtained by rewriting our general chemical reaction

$$a\text{A} + b\text{B} + \cdots \rightarrow c\text{C} + d\text{D} + \cdots$$

as

$$c\text{C} + d\text{D} + \cdots - a\text{A} - b\text{B} - \cdots = 0$$

where the reactants have been subtracted from the products. We then have

$$\text{stoichiometric number of A} = v_\text{A} = -a$$

$$\text{stoichiometric number of B} = v_\text{B} = -b$$

$$\text{stoichiometric number of C} = v_\text{C} = c$$

$$\text{stoichiometric number of D} = v_\text{D} = d$$

Worked example 2.17

Calculate the equilibrium constant for the dissociation of ethanoic acid at 25°C if the equilibrium concentrations are

$$[\text{CH}_3\text{COOH}] = 0.90 \text{ mol dm}^{-3}$$

and

$$[\text{H}^+] = [\text{CH}_3\text{COO}^-] = 0.004 \text{ mol dm}^{-3}$$

Ethanoic acid is more commonly known industrially by its older name of acetic acid. Along with acetic anhydride, it is produced by the catalytic oxidation of acetaldehyde, and is used in the production of cellulose acetate which is needed for the manufacture of upholstery and carpets.

Chemical background

Ethanoic acid is a weak electrolyte, and we can see from the concentration values above that the concentration of undissociated acid is far higher than that of the dissociated ions. We therefore expect to obtain an equilibrium constant value which is considerably less than 1. When the equilibrium constant is applied to acid dissociations such as this, it is usually given the symbol K_a. This leads to the use of logarithms for expressing acid strength in terms of pK_a, which is defined as

$$pK_a = -\log K_a$$

where log represents the logarithm taken to base 10.

Solution to worked example

The first step in this problem is to write a balanced chemical equation for the dissociation of ethanoic acid. The species required are all given in the question and the required equation is

$$CH_3COOH \rightleftharpoons CH_3COO^- + H^+$$

Subtracting the reactant from the ethanoic acid product gives

$$CH_3COO^- + H^+ - CH_3COOH = 0$$

and the stoichiometric numbers are therefore

$$\nu_{CH_3COO^-} = 1$$

$$\nu_{H^+} = 1$$

$$\nu_{CH_3COOH} = -1$$

Substituting into the general expression for the equilibrium constant then gives

$$K = \left(\frac{c_{CH_3COO^-}}{c^\ominus}\right)^1 \left(\frac{c_{H^+}}{c^\ominus}\right)^1 \left(\frac{c_{CH_3COOH}}{c^\ominus}\right)^{-1}$$

where c denotes the concentration of each species, and $c^\ominus$ is the standard concentration value of 1 mol dm^{-3}. It is acceptable to use concentrations rather than activities if we assume that the activity coefficients are all close to unity.
  Since

$$\left(\frac{c_{CH_3COOH}}{c^\ominus}\right)^{-1} = \left(\frac{1}{\frac{c_{CH_3COOH}}{c^\ominus}}\right)$$

this can be rewritten as

$$K = \frac{\left(\frac{c_{CH_3COO^-}}{c^\ominus}\right)\left(\frac{c_{H^+}}{c^\ominus}\right)}{\left(\frac{c_{CH_3COOH}}{c^\ominus}\right)}$$

The values given in the question can now be substituted into this expression:

$$K = \frac{0.004 \times 0.004}{0.90}$$

$$\simeq 1.8 \times 10^{-5}$$

Notice that the use of $c^{\ominus}$ means that the units of mol dm^{-3} do not need to appear in the equation for calculating K.

2.7.1 Quadratic equations

In section 2.2.2, we saw that a polynomial expression of order 2 is called a quadratic. A quadratic equation is one of the form

$$ax^2 + bx + c = 0$$

where a, b and c are constants. Examples are

$$3x^2 + 2x - 1 = 0$$

and

$$x^2 + 6x + 4 = 0$$

The graphs of these equations are shown in Figure 2.14, where the solutions are given by the intersection of the curves with the x-axis.

This may happen at none, one or two values of x. In some cases, these can be found by factorizing the expression into two brackets, and setting each in turn to zero. However, this is not always possible. Another method which is more generally applicable, is to apply the standard formula.

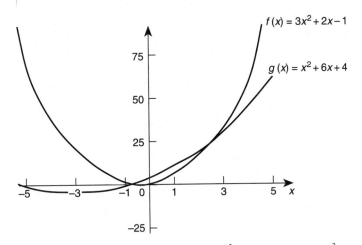

Figure 2.14 Graphs of the quadratic functions $f(x) = x^2 + 2x - 1$ and $g(x) = x^2 + 6x + 4$.

The solution of the general quadratic equation given above is

$$x = \frac{-b \pm \sqrt{b^2 - 4ac}}{2a}$$

Notice that the $\pm$ sign ensures that two values of x will result; however, solutions will only be found if b^2 is greater than the quantity $4ac$, since otherwise we would be taking the square root of a negative number. We will return to this topic again in Chapter 5.

Worked example 2.18

In the reaction between nitrogen and hydrogen to give ammonia, the fractional amount of nitrogen reacted x can be related to the equilibrium constant K which has the value 977. This gives the quadratic equation

$$81.2x^2 - 163.4x + 81.2 = 0$$

for x. What is the value of x?

Chemical background

Ammonia is also obtained commercially as a by-product of the generation of coke from coal. It is contained in an aqueous phase and is liberated by means of steam distillation. The largest use of ammonia is in fertilizers, but it is also used on a large scale in the production of nitric acid.

Ammonia is manufactured industrially by means of the Haber process. This involves the reaction

$$N_{2(g)} + 3H_{2(g)} \rightleftharpoons 2NH_{3(g)}$$

which takes place at 450°C and 200 atmospheres in the presence of an iron catalyst. These conditions are used to give an acceptable reaction rate; the equilibrium constant gives us no information about this. Nitrogen is obtained from the air while hydrogen is usually obtained from the reaction of natural gas with steam. The Haber process is actually a major use for hydrogen gas.

Solution to worked example

Relating the above expression to the general equation

$$ax^2 + bx + c = 0$$

gives $a = 81.2$, $b = -163.4$ and $c = 81.2$. Substituting these values into the equation for x

$$x = \frac{-b \pm \sqrt{b^2 - 4ac}}{2a}$$

gives

$$x = \frac{163.4 \pm \sqrt{(-163.4)^2 - (4 \times 81.2 \times 81.2)}}{2 \times 81.2}$$

This can be solved in stages using a calculator to give

$$x = \frac{163.4 \pm \sqrt{26699.6 - 26373.8}}{162.4}$$

$$= \frac{163.4 \pm \sqrt{325.8}}{162.4}$$

$$\simeq \frac{163.4 \pm 18.05}{162.4}$$

so

$$x \simeq 1.12 \text{ or } x \simeq 0.895$$

We have, as expected, obtained two values for x. However, if we look back at the question we see that only one is required. How do we know which one to take? If we look closely, we see that x is defined as the fractional amount of nitrogen reacted. Its value must therefore be less than 1 so we take $x = 0.895$ as our solution.

It is often necessary to eliminate one of the mathematical solutions in this way. Sometimes a negative and a positive value will be obtained, with the negative value obviously not making sense.

Exercises

1. Remove the brackets from the following expressions:
   (a) $xy(x^2 + y^2)$
   (b) $x(x + y)(x - y^2)$
   (c) $(x + y)(x - y)$
2. For each of the following functions:
   (i) write down the order of the expression;
   (ii) calculate $f(-3)$; and
   (iii) write down the derivative $df(x)/dx$.
   (a) $f(x) = 3x^4 + 2x^3 + 4x^2 + x + 6$
   (b) $f(x) = 8x^6 + x^5$
   (c) $f(x) = 8x^3 + 1$
3. Give the values of the following quantities:
   (a) $\log 24$
   (b) $\log 10$
   (c) $\ln 3$

4. Give the gradients and intercepts of the straight lines represented by the following equations:
   (a) $y = 6x + 3$
   (b) $3y = 5x + 4$
   (c) $x - 2y = 8$
5. If x and y are directly proportional, and x takes the value 10 when y is 30, what is the value of the proportionality constant?
6. State the relationship between x and y if they can take the following pairs of values:

x	64	32	16	8	4	2	1
y	1	2	4	8	16	32	64

7. If $f(x, y) = 3x^2y - 2xy$ calculate
   (a) $f(1, -1)$
   (b) $f(0, 2)$
   (c) $f(-2, 1)$
8. If $f(x, y) = 3x^2 + 2xy^2 + 4x^3y^2$ calculate

$$\frac{\partial f}{\partial x} \quad \text{and} \quad \frac{\partial f}{\partial y}$$

and write an expression for the differential $df(x, y)$.
9. Show that the $df(x, y)$ in Question 8 is an exact differential.
10. Solve these quadratic equations:
   (a) $x^2 + 9x - 7 = 0$
   (b) $4x^2 + 7x + 2 = 0$

Problems

1. If a gas obeys the ideal gas equation $pV = nRT$, state the value of the gradient obtained when each of the following plots were made:
   (a) p against T at constant n and V
   (b) p against n at constant V and T
   (c) V against n at constant p and T.
2. The equilibrium constant K is related to the standard enthalpy of reaction $\Delta H^{\ominus}$ and the standard entropy of reaction $\Delta S^{\ominus}$ by the equation

$$\ln K = -\left(\frac{\Delta H^{\ominus}}{RT}\right) + \left(\frac{\Delta S^{\ominus}}{R}\right)$$

where R is the ideal gas constant and T is the absolute temperature.

Calculate

$$\frac{d(\ln K)}{dT}$$

assuming that $\Delta H^{\ominus}$ and $\Delta S^{\ominus}$ are independent of temperature.

3. Derive an expression for the differential dV for a gas which obeys the ideal gas equation $pV = nRT$, assuming that n is a constant.

4. The van der Waals constants for oxygen used to be given as $a = 1.36$ atm l^2 mol^{-2} and $b = 0.0318$ l mol^{-1}. Use the conversion factors below to express these quantities in the appropriate SI units, which are Pa m^6 mol^{-2} and m^3 mol^{-1} respectively.

$$1 \text{ atm} = 101.325 \text{ kPa}$$

$$1 \text{ l} = 1 \text{ dm}^3$$

$$1 \text{ Pa} = 1 \text{ N m}^{-2}$$

5. At relatively low pressures, it is possible to ignore the constant b in the van der Waals equation to give the equation of state

$$pV^2 - RTV + a = 0$$

Solve this quadratic equation in V.

3 Solution chemistry

In some respects, the physical chemistry of liquid solutions is more complicated than that of either gases or solids. Molecules in the gas phase have relatively large separations, so the interactions between them are reduced and many approximations may be used to simplify the maths involved. Conversely, if a solid is crystalline it can be considered to be composed of many repeats of a simple unit, which again reduces the complexity of the maths required.

In this chapter, we will be concerned only with aqueous solutions, as these form by far the largest group of interest. Much of what we do will come down to unravelling the forms of the various equations so that they can be of use to us.

3.1
The Nernst equation

When an ionic substance is dissolved in water, the individual positive and negative ions each become surrounded by water molecules. This has the effect of separating the positive and negative charges which are then able to move under the influence of an electric field, as shown in Figure 3.1.

This is the basis of the electrolytic cell which can be used for the industrial production of chlorine and sodium hydroxide from sodium chloride.

If such a solution is combined with a pair of appropriate electrodes it may be possible to set up a complete electrical circuit (Figure 3.2) and the whole arrangement is said to constitute an electrochemical cell. The chemical reaction which takes place will lead to a potential difference between the electrodes; this is known as the **electromotive force** (e.m.f.) of the cell, and is usually given the symbol E.

We need to distinguish two values of E. Its value under standard conditions of unit concentration at 1 atm pressure and 25°C can be calculated from

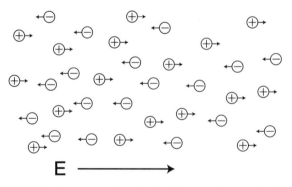

$E \longrightarrow$

Figure 3.1 Charges moving under the influence of an electric field.

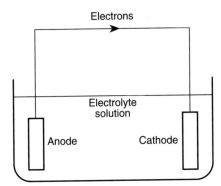

Electrons

Electrolyte
solution

Anode   Cathode

Figure 3.2 A typical electrochemical cell.

appropriate values for half reactions given in tables, and is properly called $E^{\ominus}$. When standard conditions do not prevail, the actual value is E, which can be determined from $E^{\ominus}$ using the Nernst equation. This is expressed as

$$E = E^{\ominus} - \left(\frac{RT}{zF}\right) \ln Q$$

where R is the ideal gas constant and F is the Faraday constant which has the value $96\,500$ C mol^{-1}. The quantity z can be found from the expression

$$z = \frac{n}{\text{mol}}$$

where n is the amount of electrons transferred in the reaction. Thus in a reaction involving the transfer of 2 mol of electrons, z will have a value of 2, with no units. Q is the reaction quotient, which is calculated in the same mathematical way as the equilibrium constant (see section 2.7) although the system itself is not at equilibrium. In dilute solutions, we can use concentration c in place of activity, and so

$$Q = \prod_j \left(\frac{c_j}{c^{\ominus}}\right)^{v_j}$$

where $c^{\ominus}$ is the standard concentration 1 mol dm^{-3} and v_j is the stoichiometric number of species j in the reaction equation.

A suitable source of data on half cell potentials is *Table of Standard Electrode Potentials* by G. Milazzo, S. Caroli and V. K. Sharma, published by Wiley, Chichester, 1978.

Worked example 3.1

In the reaction

$$\text{Co}_{(s)} + \text{Ni}^{2+}_{(aq)} \rightarrow \text{Co}^{2+}_{(aq)} + \text{Ni}_{(s)}$$

2 moles of electrons are transferred. If the standard electromotive potential is 0.03 V at 25°C, calculate the cell potential when $[\text{Ni}^{2+}] = 1$ mol dm^{-3} and $[\text{Co}^{2+}] = 0.1$ mol dm^{-3}.

Chemical background

The overall equation for this reaction is obtained by combining the half reactions

$$Co_{(s)} \rightarrow Co^{2+}{}_{(aq)} + 2e^-$$

and

$$Ni^{2+}{}_{(aq)} + 2e^- \rightarrow Ni_{(s)}$$

In this case, cobalt is used to liberate nickel from its salts. Metals can be arranged in an activity series, in which they would only displace those below them in the series from solution.

These show us that 2 moles of electrons are transferred from cobalt to nickel in the reaction as written but, of course, these do not appear in the overall equation.

The Nernst equation can be derived from fundamental thermodynamics, and has been shown to be valid in predicting experimental data.

Solution to worked example

The first quantity to calculate is the reaction quotient Q. We can do this by rewriting the reaction equations as products minus reactants, to give

$$Co^{2+}{}_{(aq)} + Ni_{(s)} - Co_{(s)} - Ni^{2+}{}_{(aq)} = 0$$

We therefore have the stoichiometric numbers

$$\nu_{Co^{2+}} = 1$$
$$\nu_{Ni} = 1$$
$$\nu_{Co} = -1$$
$$\nu_{Ni^{2+}} = -1$$

and so

$$Q = \left(\frac{c_{Co^{2+}}}{c^{\ominus}}\right)\left(\frac{c_{Ni}}{c^{\ominus}}\right)\left(\frac{c_{Co}}{c^{\ominus}}\right)^{-1}\left(\frac{c_{Ni^{2+}}}{c^{\ominus}}\right)^{-1}$$

which can be rewritten, since

$$x^{-1} = \frac{1}{x}$$

as

$$Q = \frac{\left(\frac{c_{Co^{2+}}}{c^{\ominus}}\right)\left(\frac{c_{Ni}}{c^{\ominus}}\right)}{\left(\frac{c_{Co}}{c^{\ominus}}\right)\left(\frac{c_{Ni^{2+}}}{c^{\ominus}}\right)}$$

Since the concentration of any solid is usually taken as constant and given the value of unity, we can set

$$c_{Ni} = 1 \text{ mol dm}^{-3} \quad \text{and} \quad c_{Co} = 1 \text{ mol dm}^{-3}$$

and so

$$\frac{c_{Ni}}{c^{\ominus}} = \frac{1 \text{ mol dm}^{-3}}{1 \text{ mol dm}^{-3}} = 1$$

Similarly,

$$\frac{c_{Co}}{c^{\ominus}} = 1$$

and so the expression for Q simplifies to

$$Q = \frac{\left(\dfrac{c_{Co^{2+}}}{c^{\ominus}}\right) \times 1}{\left(\dfrac{c_{Ni^{2+}}}{c^{\ominus}}\right) \times 1}$$

$$= \frac{c_{Co^{2+}}}{c_{Ni^{2+}}}$$

Substituting the values given in the question,

$$c_{Co^{2+}} = [Co^{2+}] = 0.1 \text{ mol dm}^{-3}$$

and

$$c_{Ni^{2+}} = [Ni^{2+}] = 1 \text{ mol dm}^{-3}$$

gives

$$Q = \frac{0.1 \text{ mol dm}^{-3}}{1 \text{ mol dm}^{-3}} = 0.1$$

The Nernst equation requires the natural logarithm of this quantity, so using a calculator gives

$$\ln Q = \ln 0.1 \simeq -2.30 \; \text{▦}$$

The next quantity to calculate is

$$\frac{RT}{zF}$$

The temperature is given as 25°C, but we require the absolute temperature which is $(25 + 273)$ K = 298 K. We are told that $n = 2$ mol, so

$$z = \frac{2 \text{ mol}}{\text{mol}} = 2$$

Combining these values with those of the constants given leads to

$$\frac{RT}{zF} = \frac{8.314 \text{ J K}^{-1}\text{ mol}^{-1} \times 298 \text{ K}}{2 \times 96\,500 \text{ C mol}^{-1}}$$

$$\simeq 0.013 \text{ J C}^{-1}$$

The units of J C^{-1} are actually the same as V, so the quantity RT/zF has the value of 0.013 V.

We are now in a position to substitute into the Nernst equation using $E^{\ominus} = 0.03$ V. This gives us

$$E = E^{\ominus} - \left(\frac{RT}{zF}\right) \ln Q$$

$$= 0.03 \text{ V} - (0.013 \text{ V} \times -2.30)$$

The two negative signs, when multiplied, give a positive quantity, so

$$E = 0.03 \text{ V} + (0.013 \text{ V} \times 2.30)$$

$$\simeq 0.03 \text{ V} + 0.03 \text{ V}$$

$$= 0.06 \text{ V}$$

3.1.1   Straight line graphs

We have already seen, in section 2.4.2, that a straight line can be represented by an equation of the form

$$y = mx + c$$

with m being the gradient of the line and c the intercept on the y-axis, as shown in Figure 3.3.

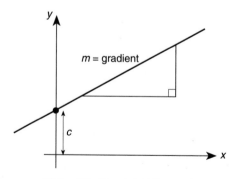

Figure 3.3   The straight line graph.

Worked example 3.2

In a series of measurements made on an electrochemical cell, the cell potential E was determined as a function of (a) reaction quotient Q and (b) temperature T. What information can be deduced from an appropriate straight line graph in each case?

Chemical background

In practice, plots of this nature are used to determine the standard cell potential $E^{\ominus}$ for the overall reaction. From such values, it is possible to assign standard electrode potentials to individual couples, which are tabulated in the literature. From these, it is possible to determine standard cell potentials by calculation rather than by measurement.

For example, the standard potentials for the reduction half reactions

$$Cu^{2+} + 2e^- \rightarrow Cu$$

and

$$Zn^{2+} + 2e^- \rightarrow Zn$$

are given as $+0.34$ V and -0.76 V respectively. If we combine these to give an overall cell potential which is positive, we need to reverse the reaction involving zinc to give

$$Zn \rightarrow Zn^{2+} + 2e^-$$

which leads to the overall cell reaction

$$Cu^{2+} + Zn \rightarrow Cu + Zn^{2+}$$

and a corresponding cell potential of

$$+0.34 \text{ V} + (+0.76 \text{ V}) = 1.10 \text{ V}$$

Solution to worked example

We need to rearrange the Nernst equation

$$E = E^{\ominus} - \left(\frac{RT}{zF}\right) \ln Q$$

into the form of the straight line graph
$$y = mx + c$$

in each case.

(a) The reaction quotient Q appears in the form of $\ln Q$ in the Nernst equation, so it seems reasonable to take E and $\ln Q$ as the variables to be plotted.

$E^\ominus$, R, T, z and F are all constants in this case. If we change the order of the terms on the right-hand side of the Nerst equation, it becomes

$$E = -\left(\frac{RT}{zF}\right)\ln Q + E^\ominus$$

Then, we can identify E with y, $-RT/zF$ with m, $\ln Q$ with x and $E^\ominus$ with c. A plot of E against $\ln Q$ will therefore have gradient $-RT/zF$ and intercept $E^\ominus$.

(b) In this case Q (and consequently $\ln Q$) remains constant while the temperature T varies. Grouping the constant terms in the rearranged Nernst equation above now gives

$$E = -\left(\frac{R\ln Q}{zF}\right)T + E^\ominus$$

so, if E is plotted against T, we have gradient $-(R\ln Q)/zF$ and intercept $E^\ominus$.

These two graphs are shown in Figure 3.4.

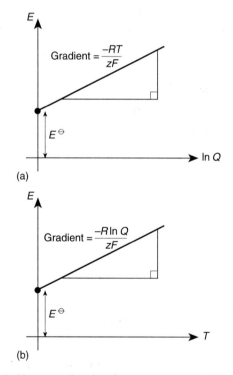

Figure 3.4 Nernst equation plot of E: (a) against $\ln Q$; and (b) against T.

The Debye–Hückel Limiting Law expresses the mean activity coefficient $\gamma_\pm$ in terms of the ionic strength I and the charges z_+ and z_- on the positive and negative ions respectively:

$$\log \gamma_\pm = -0.51 |z_+ z_-| \sqrt{\frac{I}{\text{mol dm}^{-3}}}$$

The symbol '$|x|$' simply means 'take the value of x without regard to sign'; this is called the **modulus of x**. Thus $|-5|$ would be 5, and $|5|$ would also be 5. The symbol 'log' is used to refer to logarithms to the base 10, and the reason for dividing I by units of mol dm^{-3} is to ensure that the quantity on the right of the equation has no units. This must be true on the left of the equation as the quantity is logarithmic and the logarithm of a number has no units. The mean activity coefficient $\gamma_\pm$ is defined as

$$\sqrt{\gamma_+ \gamma_-}$$

for a $1:1$ electrolyte, with γ_+ and γ_- being the activity coefficients of the positive and negative ions respectively.

The Debye–Hückel Limiting Law gives reasonable results at low concentrations, but the deviations from it as concentrations increase are considerable. The reduction in the activity coefficient of an ion is due to its interactions with the surrounding ionic atmosphere; the effect increases with charge and in solvents of low dielectric constant, as both these factors increase the magnitude of the interactions.

The relative dielectric constant is a dimensionless quantity defined as the ratio of the dielectric constant of the medium to the dielectric constant of a vacuum. Typical values of the relative dielectric constant are 78.54 for water and 2.015 for cyclohexane at 25°C.

3.2.1   Logarithms

The subject of logarithms was discussed in section 2.4.1.

Worked example 3.3

Use the Debye–Hückel Limiting Law to calculate the ionic strength of a solution of potassium chloride having a mean activity coefficient of 0.927.

Chemical background

The ionic strength I of a solution is defined by the equation

$$I = 0.5 \sum_i m_i z_i^2$$

Molality m has the units mol kg^{-1} (referring to the solvent) while **molarity** c has units of mol dm^{-3} (referring to the solution). It is preferable to use the term concentration instead of molarity to avoid confusion with molality.

where m_i is the molality of the ions i in the solution, and z_i is their charge. In dilute solutions, it is usually more convenient to replace the molalities m_i by concentrations c_i. The expression then becomes

$$I = 0.5 \sum_i c_i z_i^2$$

The rates of ionic reactions are found to vary between solutions of different ionic strength, and the second-order rate constant k is given by the equation

$$\log k = \log k_0 + 1.02 z_A z_B \sqrt{\frac{I}{\text{mol dm}^{-3}}}$$

where k_0 is a constant which depends upon the concentrations of the activated complexes present.

Solution to worked example

The first stage in this problem is to rearrange the equation which expresses the Debye–Hückel Limiting Law so that the term in ionic strength appears on its own on the left-hand side. Dividing both sides by $-0.51|z_+ z_-|$ and reversing the equation gives us the expression

$$\sqrt{\frac{I}{\text{mol dm}^{-3}}} = \frac{\log \gamma_\pm}{-0.51|z_+ z_-|}$$

In this case, it is probably easier to obtain a value for

$$\sqrt{\frac{I}{\text{mol dm}^{-3}}}$$

and then square it, rather than trying to calculate I directly. Since, in solution, potassium chloride consists of K^+ and Cl^- ions, the values of z_+ and z_- are 1 and -1 respectively. Substituting in the rearranged equation above now leads to

$$\sqrt{\frac{I}{\text{mol dm}^{-3}}} = \frac{\log 0.927}{-0.51 \times |1 \times -1|}$$

Since $|1 \times -1| = |-1| = 1$, we then have

$$\sqrt{\frac{I}{\text{mol dm}^{-3}}} = \frac{-\log 0.927}{0.51}$$

$$\simeq \frac{-(-0.0329)}{0.51}$$

$$\simeq 0.0645$$

(Note that the value of log 0.927 is negative.) To obtain the ionic strength, we now have to square both sides to give

$$\frac{I}{\text{mol dm}^{-3}} = 0.0645^2 \approx 0.0042$$

and therefore $I \approx 0.0042$ mol dm^{-3}.

3.3 Ostwald's Dilution Law

The molar conductivity Λ of an ionic solution is a measure of how well it conducts electricity, and is measured in units of Ω^{-1} cm^2 mol^{-1} or S cm^2 mol^{-1}. (Ω is the symbol for the ohm, the unit of electrical resistance and S is the symbol for siemens where 1 S $= 1$ Ω^{-1}.) It is possible to extrapolate measured values of Λ to zero concentration and obtain a value for Λ_0, the molar conductivity at infinite dilution.

The molar conductivities of K$^+$ and Cl$^-$ are 73.5 and 76.4 S cm^2 mol^{-1} respectively, at a concentration of 0.01 mol dm^{-3}. In practice, the conductivity varies with concentration.

For the dissociation of an electrolyte XY according to the equation

$$XY \rightleftharpoons X^+ + Y^-$$

the equilibrium constant K_c in terms of concentration is defined as

$$K_c = \frac{[X^+][Y^-]}{[XY]}$$

but it is also possible to define the dimensionless equilibrium constant K in terms of the molar conductivities:

$$K = \frac{c\left(\dfrac{\Lambda}{\Lambda_0}\right)^2}{1 - \left(\dfrac{\Lambda}{\Lambda_0}\right)}$$

which is a satisfactory expression for many weak electrolytes.

3.3.1 Discontinuities

A function has a **discontinuity** when it is not defined for every possible input value. For example, consider the function defined by

$$f(x) = \frac{x + 1}{x - 4}$$

For a function to be defined, there has to be a single value that results from evaluating the function for a given value of x. If we evaluate the function f for the first few integers, we obtain

$$f(0) = \frac{1}{-4} = -\tfrac{1}{4}$$

$$f(1) = \frac{2}{-3} = -\tfrac{2}{3}$$

$$f(2) = \frac{3}{-2} = -\tfrac{3}{2}$$

$$f(3) = \frac{4}{-1} = -4$$

$$f(4) = \frac{5}{0}$$

When $x = 4$, the denominator of $f(x)$ is zero. Division by zero results in an infinite number of possible answers, none of which are acceptable, so $f(x)$ is not defined when $x = 4$. A plot of the function $f(x)$ is shown in Figure 3.5. Notice that if the value of x is very close to 4 then the function is still defined; it is discontinuous *only* at the exact value $x = 4$.

Worked example 3.4

The expression of Ostwald's Dilution Law above is an example of a function which has a discontinuity. In this case, it is not possible to obtain a value of K for every value of Λ. For what value or values of Λ is K undefined?

Chemical background

For strong electrolytes, the concept of degree of dissociation does not apply. If they are virtually fully ionized the degree of dissociation α will be equal to 1,

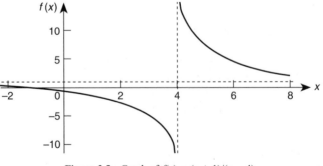

Figure 3.5   Graph of $f(x) = (x + 1)/(x - 4)$.

and the properties of their solutions will depend to a large extent on the nature of the solvent.

Solution to worked example

In the equation

$$K = \frac{c\left(\dfrac{\Lambda}{\Lambda_0}\right)^2}{1 - \left(\dfrac{\Lambda}{\Lambda_0}\right)}$$

K will be undefined when the denominator is zero, i.e. when

$$1 - \frac{\Lambda}{\Lambda_0} = 0$$

$$\frac{\Lambda}{\Lambda_0} = 1$$

It follows that K is not defined when $\Lambda = \Lambda_0$.

For an electrolyte MX, the degree of dissociation α is defined as

$$\alpha = \frac{m_{X^-}}{m}$$

where m_{X^-} is the molality of the dissociated anion and m is the overall molality of the MX solution.

**3.4
Fick's laws**

If we were to take equal volumes of two solutions of, say, sodium chloride in water at different concentrations and mix them, we would expect to find that the resulting solution had a concentration between that of the two starting solutions (Figure 3.6). This is because molecules of sodium chloride will move from the more concentrated solution to the less concentrated solution. We say that there has been a diffusion of sodium chloride molecules under the influence of a concentration gradient.

The concentration gradient is simply the change in concentration c over a given distance. This may be a constant, in which case it is straightforward to calculate, or it may need to be defined as a derivative such as dc/dx where x simply represents the direction in which the concentration varies. In general, we deal with three-dimensional problems, so it is possible to define a concentration

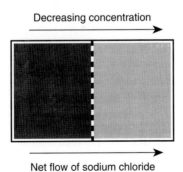

Decreasing concentration

Net flow of sodium chloride

Figure 3.6   Diffusion of molecules under the influence of a concentration gradient.

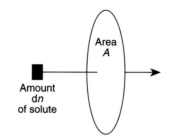

Figure 3.7   Definition of the diffusive flux, *J*.

gradient in the direction of each of the *x*, *y* and *z* axes. We now need to specify partial derivatives, since we are effectively varying only one of the three possible variables *x*, *y* and *z*. The possible concentration gradients will then be $\partial c/\partial x$, $\partial c/\partial y$ and $\partial c/\partial z$.

The diffusive flux, *J*, is the rate at which a solute crosses an area *A*, and is defined by the derivative dn/dt where an amount dn of solute crosses the area in time dt (Figure 3.7). Fick's first law relates *J* to the concentration gradient, and is expressed by the equation

$$J = -DA \frac{\partial c}{\partial x}$$

where *D* is the diffusion coefficient.

The diffusion coefficient for the diffusion of one gas into another may be determined using a cell with a sliding partition. This is removed for a short interval of time which is measured. The compositions of each chamber are determined before and after the diffusion takes place.

3.4.1   Proportion

The subject of proportion was discussed in section 2.5.3.

Worked example 3.5

25 cm^3 of a solution of sodium chloride of concentration 1.0 mol dm^{-3} was mixed with 75 cm^3 of a solution of sodium chloride of concentration 0.1 mol dm^{-3}. What is the concentration of the resulting solution?

Chemical background

It is quite reasonable to solve this problem by first calculating the amount of sodium chloride in each solution, then the total in the final solution, and dividing by the new volume to obtain the concentration. However, applying the ideas of proportion gives us a useful alternative which may be quicker and can also be used to verify the alternative.

You should note that an extreme case of diffusion along a concentration gradient is that of dilution. Here, the lower concentration is initially zero.

The tendency for molecules to diffuse can be predicted from thermodynamics, as they seek to attain a state which has higher entropy.

Solution to worked example

The total volume of solutions being mixed is

$$(25 + 75) \text{ cm}^3 = 100 \text{ cm}^3$$

The contributions to the overall solution are therefore

$$\frac{25}{100} \text{ and } \frac{75}{100}$$

for the 1.0 and 0.1 mol dm^{-3} solutions respectively. Since the contribution from each will increase with their concentrations, it will be *directly* proportional to their concentrations.

The contribution from 1.0 mol dm^{-3} solution will be

$$\frac{25}{100} \times 1.0 \text{ mol dm}^{-3} = 0.25 \text{ mol dm}^{-3}$$

The contribution from 0.1 mol dm^{-3} solution will be

$$\frac{75}{100} \times 0.1 \text{ mol dm}^{-3} = 0.075 \text{ mol dm}^{-3}$$

The total concentration is now found by adding these two contributions, which gives a value of 0.325 mol dm^{-3}. (The inclusion of 3 significant figures assumes, of course, that all the figures given in the question were exact!)

Worked example 3.6

The net rate of increase of concentration c with time t is given by the expression

$$\frac{\partial c}{\partial t} = -\frac{1}{A}\frac{\partial J}{\partial x}$$

where J is the rate of diffusion across an area A. Use Fick's first law to eliminate J from this equation, assuming that the diffusion coefficient D is independent of distance x.

Chemical background

The expression to be obtained in this problem is actually a statement of Fick's second law of diffusion, and is also known as the **diffusion equation**. It applies when diffusion takes place in one direction, denoted arbitrarily as x in this case. The diffusion coefficient D is the same in all directions in isotropic media only. The equation is actually of more general applicability than the diffusion of particles, and can be applied to the variation of properties such as temperature.

The equation can be solved by realizing that the diffusive flux J must be zero at either end of the concentration gradient. This is an example of a boundary condition, which is a concept we will meet again in Chapter 6 when we discuss the subject of quantum mechanics. A full solution of the diffusion equation leads to the expression

$$c = \frac{n_0}{A(\pi Dt)^{1/2}} \exp\left(\frac{x^2}{4Dt}\right)$$

where n_0 is the total (and constant) amount of solute, and exp refers to the exponential function (see section 4.6.1).

Solution to worked example

It can be difficult to know where to start in solving problems like this. The important thing you need to realize is that we are asked to eliminate J from the given equation, in which it appears as the partial derivative $\partial J/\partial x$. We therefore need to find some way of obtaining an expression for this partial derivative. Since Fick's first law gives J as a function of x, this should be possible. If

$$J = -DA\frac{\partial c}{\partial x}$$

as given earlier, it follows that

$$\frac{\partial c}{\partial t} = -\frac{1}{A}\frac{\partial J}{\partial x}$$

can be rewritten as

$$\frac{\partial c}{\partial t} = -\frac{1}{A}\frac{\partial}{\partial x}\left(-DA\frac{\partial c}{\partial x}\right)$$

We are told that D is independent of x, so it can be treated as a constant and brought outside the derivative. Similarly, A is a constant and does not need to be included in the derivative. Noticing also that multiplying two negative values gives a positive value, this leaves us with

$$\frac{\partial c}{\partial t} = \frac{DA}{A}\frac{\partial}{\partial x}\left(\frac{\partial c}{\partial x}\right)$$

in which the As will cancel, to give

$$\frac{\partial c}{\partial t} = D \frac{\partial}{\partial x}\left(\frac{\partial c}{\partial x}\right)$$

We now have the partial derivative

$$\frac{\partial}{\partial x}\left(\frac{\partial c}{\partial x}\right)$$

which denotes differentiation of c once with respect to x, then once more with respect to x. In total then, c needs to be differentiated twice with respect to x, giving us what is known as the **second-order partial derivative** $\partial^2 c/\partial x^2$.

The final expression therefore becomes

$$\frac{\partial c}{\partial t} = D \frac{\partial^2 c}{\partial x^2}$$

3.5
Density of solutions

The density of a solution can be calculated from the simple formula

$$\text{density} = \frac{\text{mass}}{\text{volume}}$$

for a given solution. The density of a solution containing a given solute and solvent will also vary according to the mole fraction of solute. An example of this is that the density $\rho(x)$ of a solution of a certain alcohol in water is described by the function

$$\frac{\rho(x)}{\text{g cm}^{-3}} = 0.982 - 0.265x + 0.306x^2$$

at a temperature of 25°C, with x being the mole fraction of alcohol in the water. This function is shown in Figure 3.8.

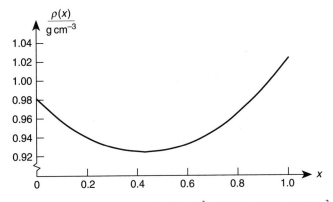

Figure 3.8 Graph of the function $\rho(x)/\text{g cm}^{-3} = 0.982 - 0.265x + 0.306x^2$.

It is then possible to evaluate the density for any solution having a specified value of x, and to determine the maximum and minimum values of the density for such solutions.

3.5.1   Functions

The subject of functions was discussed in section 2.2.3.

Worked example 3.7

Such a solution will be extensively hydrogen-bonded, resulting in a high solubility of alcohol in water.

Calculate the density of the alcohol–water solution which obeys the equation above and has a mole fraction of alcohol of 0.05.

Chemical background

The measurement of density, from the basic defining equation, requires the accurate measurement of the mass of a known volume of substance. For liquids and solutions, such measurements can be achieved by using a pycnometer, such as that shown in Figure 3.9. The pycnometer needs to be completely dry before being filled to the very top, and after the stopper is replaced it needs to be carefully wiped to remove the displaced liquid. Before weighing, the filled pycnometer should be immersed in a thermostated bath at the required temperature.

Solution to worked example

This problem requires a straightforward substitution of values into the equation given above. Using the function notation gives us

$$\frac{p(0.05)}{\text{g cm}^{-3}} = 0.982 - (0.265 \times 0.05) + (0.306 \times 0.05^2)$$

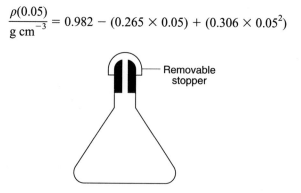

Removable stopper

Figure 3.9   A pycnometer.

Evaluating each of these terms using a calculator gives

$$0.265 \times 0.05 = 0.013\,25$$
$$0.306 \times 0.05^2 = 7.65 \times 10^{-4}$$

so

$$\frac{p(0.05)}{\text{g cm}^{-3}} = 0.982 - 0.013\,25 + 7.65 \times 10^{-4}$$

Since the first term in the expression is given to 3 decimal places, we are not justified in giving the final answer to any more places. We do however need to consider terms to one more decimal place to allow for rounding. This gives us

$$\frac{p(0.05)}{\text{g cm}^{-3}} \approx 0.9695 \approx 0.970$$

so the final value for the density is 0.970 g cm^{-3}.

3.5.2 Stationary points

We saw in section 2.3 that we could think of the derivative dy/dx as being the gradient of the graph of y against x, which could also be found by drawing the tangent to the curve at the point of interest. Suppose that we have a function $f(x)$ whose graph is shown in Figure 3.10. This has a maximum (peak) and a minimum (trough) at the points indicated; these are known as **stationary points**. Notice that at both the maximum and the minimum the tangent to the curve is horizontal, and consequently has a value of zero.

This fact gives us a way of determining the points at which a function $f(x)$ has a stationary point. If the tangent to the curve is zero, then

$$\frac{df}{dx} = 0$$

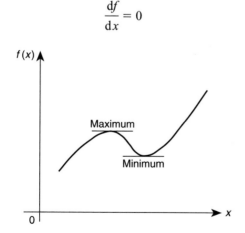

Figure 3.10   Graph of a function which has a maximum and a minimum.

It is also possible to identify the nature of the stationary points found. This is done by looking at the value of the second derivative d^2f/dx^2.

Value of $\dfrac{d^2f}{dx^2}$	Type of stationary point
Negative	Maximum
Zero	Point of inflexion
Positive	Minimum

A point of inflexion, as in the middle of the letter S, is both a maximum and minimum stationary point.

Worked example 3.8

For the solution referred to in Worked Example 3.7 determine the value of the alcohol mole fraction for which the solution density is a maximum or a minimum. Identify the nature of the stationary point.

Chemical background

Expressions such as this for the density of a solution can be used to obtain expressions for the partial molar volume, which we will meet in the next section. In general, the more terms which are included in expressions of this type, the better the fit to experimental data. There are examples of this type of relationship which contain terms up to x^5, but these are difficult to deal with mathematically. In such a case the derivative $d\rho/dx$ would contain terms up to x^4. Polynomials including terms to x^4 are called **quartic**. Setting $d\rho/dx$ equal to zero would result in the need to solve a quartic equation in x. We often use approximations in physical chemistry to simplify such expressions so that we can obtain meaningful results. Since x is less than 1 in this case, the higher terms in x become successively smaller, and because the coefficients of these terms are all of similar magnitude, these higher terms can be neglected.

Solution to worked example

To determine the maxima and minima, we need to identify the values of x for which

$$\frac{d\rho(x)}{dx} = 0$$

Notice that the expression for $\rho(x)$ contains terms in x and x^2 as well as a constant term, so we need to use the following standard derivatives:

$$\frac{d(\text{constant})}{dx} = 0$$

$$\frac{d(x)}{dx} = 1$$

$$\frac{d(x^2)}{dx} = 2x$$

Differentiating

$$\rho(x) = 0.982 - 0.265x + 0.306x^2$$

now gives us

$$\frac{d\rho(x)}{dx} = 0 - (0.265 \times 1) + (0.306 \times 2x)$$

which can be tidied up with the aid of a calculator to give

$$\frac{d\rho(x)}{dx} = -0.265 + 0.612x$$

We have already seen that stationary points will arise when

$$\frac{d\rho(x)}{dx} = 0$$

so we need to solve the equation

$$-0.265 + 0.612x = 0$$

This is a linear equation which can be solved by a simple rearrangement. The first step is to isolate the term in x on the left of the equation. To do this, we add 0.265 to both sides of the equation, and this gives

$$0.612x = 0.265$$

Now dividing each side by 0.612, to leave x on the left gives

$$x = \frac{0.265}{0.612} \simeq 0.433$$

This value of x can be substituted into the formula for $\rho(x)$ to give

$$\frac{\rho(0.433)}{\text{g cm}^{-3}} = 0.982 - (0.265 \times 0.433) + (0.306 \times 0.433^2)$$

$$\simeq 0.982 - 0.1147 + 0.0574$$

$$\simeq 0.925$$

so that $\rho = 0.925$ g cm^{-3}. As in Worked Example 3.7, the answer is given to 3 decimal places, which is consistent with the original data.

We now need to determine the nature of this stationary point. We saw above that

$$\frac{d\rho}{dx} = -0.265 + 0.612x$$

The second derivative, $d^2\rho/dx^2$ or

$$\frac{d}{dx}\left(\frac{d\rho}{dx}\right)$$

is then found by differentiating again. We need to use the standard derivatives:

$$\frac{d}{dx}(\text{constant}) = 0$$

$$\frac{d}{dx}(x) = 1$$

Then

$$\frac{d^2\rho}{dx^2} = 0.612$$

Note that $d^2\rho/dx^2$ is a constant since it does not contain any terms in x. In general, you would need to substitute the value of x found for the stationary point into our expression for $d^2\rho/dx^2$ to discover whether it was positive or negative (minimum or maximum).

In this case, this extra calculation is not necessary because ρ is a quadratic function and differentiating a quadratic twice will always result in a constant value. The positive (constant) value of $d^2\rho/dx^2$ denotes that $\rho(x)$ has a minimum when $x = 0.433$.

3.6
Partial molar volumes

When two solutions are mixed, the total volume of the resulting solution is not necessarily the sum of the individual volumes. In fact, the total volume V of solution can be expressed by the equation

$$V = n_1 V_1 + n_2 V_2$$

Typical values of partial molar volumes are 16.98 cm^3 mol^{-1} and 57.60 cm^3 mol^{-1} for water and ethanol respectively in a solution containing 0.42 mole fraction of ethanol.

where n_1 and n_2 are the amounts of components 1 and 2, and V_1 and V_2 are the corresponding respective partial molar volumes. The partial molar volume is actually defined as a partial derivative:

$$V_1 = \left(\frac{\partial V}{\partial n_1}\right)_{T, p, n_2, \ldots}$$

The subscript $T, p, n_2, \ldots$ indicates that the temperature, pressure and amount of component 2 are all held constant as we consider the variation of the volume V

with the amount n_1 of component 1. The dots allow for solutions composed of more than two components, and represent n_3, n_4, etc.

3.6.1 Functions

The subject of functions was discussed in section 2.2.3.

Worked example 3.9

The partial molar volume V_1 of water in a solution of potassium sulphate is given as a function of molality m by the equation

$$\frac{V_1}{cm^3} = -0.1094\left(\frac{m}{m^\ominus}\right)^{3/2} - 0.0002\left(\frac{m}{m^\ominus}\right)^2 + 17.963$$

where $m^\ominus$ is the standard molality of 1 mol kg^{-1}.
  Calculate the partial molar volume of water when $m = 0.10$ mol kg^{-1}.

Chemical background

Although partial molar volumes are defined in terms of the amounts n_1 and n_2 of solvent and solute, for practical purposes we often work in terms of the molality m of the solute. This is defined as the amount of solute per kg of solvent, so if we actually have (or can assume) 1 kg of solvent then m will be the same as n_2.

The conversion from molarity to molality can be made so long as we know the density of the solution.

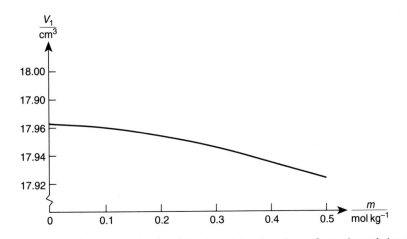

Figure 3.11 Graph of the function describing the partial molar volume of water in a solution of potassium sulphate.

Solution to worked example

The expression for V_1 can be written using the function notation as

$$\left(\frac{V_1}{cm^3}\right)\left(\frac{m}{m^\ominus}\right) = -0.1094\left(\frac{m}{m^\ominus}\right)^{3/2} - 0.0002\left(\frac{m}{m^\ominus}\right)^2 + 17.963$$

and its graph is shown in Figure 3.11. We can calculate V_1 for 0.10 mol kg^{-1} by direct substitution into the function expression. This gives us

$$\frac{V_1(0.1)}{cm^3} = -0.1094 \times 0.10^{3/2} - 0.0002 \times 0.10^2 + 17.963 \;\blacksquare$$

Evaluating this expression term by term using a calculator, gives us

$$\frac{V_1(0.10)}{cm^3} = -0.1094 \times 0.03162 - 2 \times 10^{-6} + 17.963$$

$$\approx -0.00346 - (2 \times 10^{-6}) + 17.963$$

The second term of 2×10^{-6} is much smaller than the other two and can be safely neglected because the final answer is given to 3 decimal places. We are now left with

$$\frac{V_1(0.10)}{cm^3} = -0.00346 + 17.963$$

$$= 17.95954$$

We are therefore restricted to three decimal places in our answer which will be 17.960 cm^3.

3.6.2 Stationary points

The subject of stationary points was discussed in section 3.5.2.

Worked example 3.10

The total volume V of a solution of magnesium sulphate is given in terms of its molality m as

$$\frac{V}{cm^3} = 1001.38 - 4.86\left(\frac{m}{m^\ominus}\right) + 34.69\left(\frac{m}{m^\ominus}\right)^2$$

where $m^\ominus$ is the standard molality of 1 mol kg^{-1}. For what value of the molality will the volume be at a minimum?

Chemical background

Logically, we might expect the volume to be at a minimum when no magnesium sulphate is present, and then to increase steadily as more salt is added until the solution becomes saturated. In fact, the partial molar volume of magnesium sulphate extrapolated to zero concentration has a negative value. The hydration of the Mg^{2+} and SO_4^{2-} ions causes the structure of water to break down and collapse.

Solution to worked example

To obtain the value of m when V will be a minimum, we need to solve the equation

$$\frac{dV}{dm} = 0$$

To differentiate the expression for V, we need to recall the standard rules

$$\frac{d}{dx}(x) = 1$$

$$\frac{d}{dx}(x^2) = 2x$$

The derivative is then given by

$$\frac{dV}{dm} = 0 - (4.86 \times 1) + (34.69 \times 2m)$$

$$= -4.86 + 69.38m$$

Setting dV/dm to equal zero will allow us to find the value of m at the stationary point:

$$-4.86 + 69.38m = 0$$

To solve this equation we add 4.86 to both sides to give

$$69.38m = 4.86$$

and then divide both sides by 69.38:

$$m = \frac{4.86}{69.38} \simeq 0.070 \text{ mol kg}^{-1}$$

It would be easy to think that this was the final answer to the question; however, we have not yet shown whether this value gives a maximum or a minimum value of V. Remember that we can do this by looking at the value of d^2V/dm^2 which we obtain by differentiating our expression for dV/dm once more:

$$\frac{d^2V}{dm^2} = \frac{d}{dm}\left(\frac{dV}{dm}\right)$$

$$= \frac{d}{dm}(-4.86 + 69.38m)$$

$$= 0 + (69.38 \times 1)$$

$$= 69.38$$

Again, no substitution of the value of m is required. Since 69.38 is a positive value, $m = 0.070$ mol kg^{-1} gives V its minimum value.

Exercises

1. If p is defined as

$$p = 3 - 2\ln x$$

where $x = y/z$, evaluate p when $y = 7.5$ and $z = 2.8$.
2. If r is directly proportional to both x and y and inversely proportional to z^2, obtain a general equation which relates r to x, y and z.
3. Solve the equation

$$\log_x 81 = 3$$

for x.
4. For the following equations, state which plot would give a straight line, and what the resulting values of the gradient and intercept would be.
   (a) $y = 4 + 6\ln x$
   (b) $\log y = 2\sqrt{x}$
5. At what values of x do the following functions of x have discontinuities?

   (a) $f(x) = \dfrac{x}{x^2 - 9}$

   (b) $f(x) = \dfrac{3x + 1}{2x + 3}$

6. At what values of x and/or y do the following functions of x and y have discontinuities?

   (a) $f(x, y) = \dfrac{x + y + 1}{x^2 + y^2}$

   (b) $f(x, y) = \dfrac{2y}{x(x - 1)}$

7. The function $h(x, y)$ is defined as

$$h(x, y) = \frac{x(y - 1)}{y(x - 1)}$$

and has discontinuities along two lines. Give the value of x which defines one line, and the value of y which defines the other line.

8. Locate the stationary points of the function

$$f(x) = 2x^3 - 24x + 7$$

9. Locate and identify the stationary points of the function

$$g(x) = 3x^2 - 2x + 1$$

10. Locate the stationary points of the function

$$h(x) = x^2 - \sqrt{x}$$

Problems

1. For each of the following equations, state which quantities you would plot to give a straight line, and the values of the resulting gradient and intercept:

   (a) $\pi = cRT$

   Here π is the osmotic pressure which varies according to concentration c, R is the ideal gas constant and T is the specified absolute temperature.

   (b) $\log k = \log k_0 + 1.02 z_A z_B \sqrt{I}$

   Here k is the rate constant for a reaction which varies according to the ionic strength I. The rate constant for the limit of unit activity coefficients is k_0 and z_A and z_B are the charges on ions A and B respectively.

   (c) $\Lambda = \Lambda_0 - (P + Q\Lambda_0)\sqrt{c}$

   Here Λ is the molar conductivity of an electrolyte solution of concentration c, Λ_0 is the molar conductivity at infinite dilution and P and Q are constants.

2. (a) The molar mass M of a solute is given in terms of the sedimentation coefficient s and the diffusion coefficient D:

$$M = \frac{RTs}{D(1 - V_1\rho)}$$

   where R is the ideal gas constant, T the absolute temperature and V_1 the volume per unit mass of the solute. When will it *not* be possible to calculate M using this equation?

(b) For a solute which undergoes a dimerization reaction with equilibrium constant K, it can be shown that

$$K = \frac{K_b(K_b m - \Delta T_b)}{(2\,\Delta T_b - K_b m)^2}$$

where ΔT_b is the boiling point elevation, K_b the ebullioscopic constant and m the molality. Under what conditions is K undefined according to this equation?

3. (a) If v is the total number of ions resulting from the complete dissociation of a molecule and i is the van't Hoff factor related to the colligative properties of electrolytes, then the degree of dissociation α is given by the equation

$$\alpha = \frac{i - 1}{v - 1}$$

By considering possible values of v, decide whether α is a continuous function for electrolytes.

(b) For a single $1:1$ electrolyte, the dissociation constant K is given by the expression

$$K = \frac{c^2 \alpha^2}{1 - \alpha}$$

where c is the concentration and α is the degree of dissociation. Under what circumstances is K not defined?

4. The total volume V of a sodium chloride solution at 25°C is given by the expression

$$\frac{V}{\text{cm}^3} = 1003.0 + 16.4\left(\frac{n}{\text{mol}}\right) + 2.1\left(\frac{n}{\text{mol}}\right)^{3/2} + 0.003\left(\frac{n}{\text{mol}}\right)^{5/2}$$

where n is the amount of salt in 1 kg of water. Calculate the partial molar volume V_{NaCl}, which is equal to the derivative $\partial V/\partial n$ when $n = 0.15$ mol kg^{-1}.

5. The volume V of a solution of an alcohol in water is given by the equation

$$\frac{V}{\text{cm}^3} = 1004.08 + 3.9062\left(\frac{m}{m^\ominus}\right) - 2.4362\left(\frac{m}{m^\ominus}\right)^2$$

where m is the molality and $m^\ominus = 1$ mol kg^{-1}. Determine the maximum volume of this solution.

Kinetics 4

In contrast to the thermodynamics discussed in Chapter 3, the range of mathematical techniques required for a study of kinetics is somewhat smaller. We will make much use of a technique called integration, which is essentially the reverse of the differentiation process we have already met. Analysing the rates of reactions normally involves taking a derivative, which describes the rate of change of concentration with time, and then reversing the differentiation process to obtain the original function.

4.1
Rates of change

We saw in section 2.3.1 that we can consider both average and instantaneous rates of change. For chemical reactions, we normally monitor the change in concentration, or a related quantity, of the reactants or products with time. If we know the initial concentration, and its value after a specified time interval, it is possible to calculate the average rate of reaction.

Worked example 4.1

In the reaction between propionaldehyde and hydrocyanic acid, the concentration of the acid was 0.0902 mol dm^{-3} after 5.2 min and 0.1652 mol dm^{-3} after 20.0 min. What is the average rate of reaction?

Propionaldehyde is prepared from ethene, hydrogen and carbon monoxide by means of an homogeneously catalysed hydroformylation reaction.

Chemical background

Full details of this study are given in the *Journal of the American Chemical Society*, **75**, 3106, 1953. The effect of variations in the ionic strength, pH and buffer solution on the reaction rate were monitored in acetate buffers. The reaction was found to be second order.

Solution to worked example

We need to divide the change in concentration by the change in time, remembering that the change is calculated as final value minus initial value for both quantities

$$\text{change in concentration} = (0.1652 - 0.0902)\ \text{mol dm}^{-3}$$

$$= 0.0750\ \text{mol dm}^{-3}$$

$$\text{change in time} = (20.0 - 5.2)\ \text{min}$$

$$= 14.8\ \text{min}$$

So,

$$\text{average rate of reaction} = \frac{0.0750\ \text{mol dm}^{-3}}{14.8\ \text{min}}$$

$$\approx 5.07 \times 10^{-3}\ \text{mol dm}^{-3}\ \text{min}^{-1}$$

This is a perfectly correct way of presenting the answer. However, it is often useful to express such quantities in terms of SI units. In this case, the SI unit of time is seconds (s), and $60\ \text{s} = 1\ \text{min}$. Substituting this information into our answer gives

$$\text{average rate of reaction} = 5.06 \times 10^{-3}\ \text{mol dm}^{-3}\ (60\ \text{s})^{-1}$$

If we apply the power of -1 to both the number and unit in the bracket we get

$$\text{average rate of reaction} = 5.06 \times 10^{-3}\ \text{mol dm}^{-3} \times 60^{-1}\ \text{s}^{-1}$$

Now, a power of -1 is equivalent to calculating a reciprocal ('one over a quantity')

$$60^{-1} = \frac{1}{60}$$

and so

$$\text{average rate of reaction} = \frac{5.06 \times 10^{-3}\ \text{mol dm}^{-3}\ \text{s}^{-1}}{60}$$

$$\approx 8.43 \times 10^{-5}\ \text{mol dm}^{-3}\ \text{s}^{-1}$$

Worked example 4.2

In the reaction

$$H_{2(g)} + Br_{2(g)} \rightarrow 2HBr_{(g)}$$

the initial bromine concentration of $0.321\ \text{mol dm}^{-3}$ falls to a value of $0.319\ \text{mol dm}^{-3}$ after $0.005\ \text{s}$. Assuming that this time interval is small compared to the time taken for the reaction to reach completion, determine expressions for the rate of change of concentration of each chemical species with respect to time.

Chemical background

Once a complex reaction is underway, it is possible for subsidiary reactions to occur between the products and reactants. Because of this, the system may only be completely defined in terms of the species present and their states at the start of the reaction, and measurements of rate will then be more reliable than later on.

It is, of course, necessary to be able to make measurements on a time-scale which is short relative to that of the overall reaction.

Solution to worked example

We saw in section 2.3.1 that the value of $df(x)/dx$ could be considered to be the gradient of the line joining two points which had very similar values of x. Since we are told that the time interval of 0.005 s is very short in this case, the rate of change of bromine can be taken to be its instantaneous value and can therefore be represented by the derivative

$$\frac{d[Br_2]}{dt}$$

The square brackets are used to denote concentration and the subscript denoting the state is omitted for clarity.

To obtain the rate of change, we need to divide the increase in concentration by the increase in time, remembering that we always subtract initial values from final.

$$\text{change in concentration} = (0.319 - 0.321) \text{ mol dm}^{-3}$$
$$= -0.002 \text{ mol dm}^{-3}$$

Since we have defined the change in concentration as an increase, the negative value indicates that the concentration actually decreases, as we can see in the question. Now

$$\text{change in time} = 0.005 \text{ s}$$

so the instantaneous rate of change is given by the expression

$$\frac{d[Br_2]}{dt} = \frac{-0.002 \text{ mol dm}^{-3}}{0.005 \text{ s}}$$
$$= -0.4 \text{ mol dm}^{-3} \text{ s}^{-1}$$

We also need to obtain expressions for the rates of change of H_2 and HBr. Since hydrogen and bromine react in a 1 : 1 ratio, they are being consumed at the same rate, so we can write

$$\frac{d[H_2]}{dt} = -0.4 \text{ mol dm}^{-3} \text{ s}^{-1}$$

The equation also shows us that for every molecule of bromine consumed, two molecules of hydrogen bromide are produced. Therefore, while the concentration of bromine decreases that of hydrogen bromide increases. Since $d[Br_2]/dt$ has a negative value it follows that $d[HBr]/dt$ will have a positive value which is twice the magnitude:

$$\frac{d[HBr]}{dt} = -2\frac{d[Br_2]}{dt}$$

$$= -2 \times (-0.4 \text{ mol dm}^{-3} \text{ s}^{-1})$$

$$= 0.8 \text{ mol dm}^{-3} \text{ s}^{-1}$$

4.2
The order of a reaction

The relationship of the rate of a reaction to the concentrations of its reactants has to be determined experimentally. In some cases, it can be expressed by an equation of the form

$$\text{rate} = k[A]^x[B]^y \cdots$$

where A and B are reactants and x and y are indices (powers) which are determined experimentally. The constant k is known as the **rate constant**. We say that the order of the reaction with respect to A is x, and with respect to B is y. The overall order is the sum of all the indices, in this case $x + y$.

4.2.1   Indices

The index (plural indices) of a variable is simply the value of the power to which it is raised. We would say that x^5 has an index of 5 and $[Cl^-]^2$ an index of 2. There are two values of indices which can sometimes be confusing, and may need to be treated in a special way.

A variable on its own can be considered as having an index of 1, so that

$$x = x^1$$

$$[OH^-] = [OH^-]^1$$

Any number raised to the power zero is one. For example

$$x^0 = 1$$

$$[H_2]^0 = 1$$

It is sometimes useful to multiply a quantity by an appropriate variable raised to a zero power, as this does not change the value of an expression but may assist the analysis. The polynomial

$$3x^2 + 2x + 5$$

could be rewritten as

$$3x^2 + 2x^1 + 5x^0$$

for instance.

Worked example 4.3

Give the order with respect to each reactant and the overall order of the following reactions which obey the given rate laws:

(a) $H_2 + I_2 \rightleftharpoons 2HI$

$$-\frac{d[I_2]}{dt} = k[H_2][I_2]$$

(b) $BrO_3^- + 5Br^- + 6H^+ \rightleftharpoons 3Br_2 + 3H_2O$

$$-\frac{d[BrO_3^-]}{dt} = k[BrO_3^-][Br^-][H^+]^2$$

(c) $(CH_3)_3CCl + H_2O \rightleftharpoons (CH_3)_3COH + HCl$

$$-\frac{d[(CH_3)_3CCl]}{dt} = k[(CH_3)_3CCl]$$

Chemical background

(a) We met this reaction in Worked Example 1.8, which was concerned with the equilibrium constant K. This value tells us to what extent a reaction will take place, but we need to study the kinetics to determine how fast the reaction would be. Remember that we use a large (capital) K to denote an equilibrium constant, but a small (lowercase) k for a rate constant.

(b) In alkaline solutions, bromine disproportionates spontaneously to BrO^- and Br^-; the former further disproportionates to Br^- and BrO_3^-. In acidic solutions, however, the reaction proceeds spontaneously from left to right as written.

> Disproportionation occurs when more than one product is produced from a single reactant. Simultaneous oxidation and reduction then take place.

(c) The reaction between *tert*-butyl chloride and water to give *tert*-butyl alcohol is an example of nucleophilic substitution. The first-order kinetics for this reaction are a consequence of it being an S_N1 reaction, with the slowest or rate-determining step being the formation of a carbonium ion:

$$(CH_3)_3CCl \rightarrow (CH_3)_3C^+ + Cl^-$$

> Common protic solvents in which nucleophilic substitution reactions can take place are water, alcohols and carboxylic acids.

Solution to worked example

(a) The concentrations of the reactants appear as $[H_2]$ and $[I_2]$, each of which are raised to the power of one (which does not need to be written). Therefore

$$\text{order with respect to } H_2 = 1$$
$$\text{order with respect to } I_2 = 1$$
$$\text{overall order} = 1 + 1 = 2$$

(b) The concentrations of the reactants appears as $[BrO_3^-]$, $[Br^-]$ and $[H^+]^2$. If we realize that the absence of a written power implies a value of one, we have

$$\text{order with respect to } BrO_3^- = 1$$

$$\text{order with respect to } Br^- = 1$$

$$\text{order with respect to } H^+ = 2$$

$$\text{overall order} = 1 + 1 + 2 = 4$$

(c) The concentration of $(CH_3)_3CCl$ appears as $[(CH_3)_3CCl]$, which implies a power of one. Notice that the second reactant H_2O does not appear in the equation. However, since any number raised to a zero power is one, we could rewrite the rate equation as

$$-\frac{d[(CH_3)_3CCl]}{dt} = k[(CH_3)_3CCl]^1[H_2O]^0$$

We now simply read the powers to which the concentration of each reactant is raised from the equation

$$\text{order with respect to } (CH_3)_3CCl = 1$$

$$\text{order with respect to } H_2O = 0$$

$$\text{overall order} = 1 + 0 = 1$$

4.3
Zero-order reactions

From our definition in section 4.2, a zero-order reaction is one for which the sum of the powers in the rate equation is zero. Taking our general expression

$$\text{rate} = k[A]^x[B]^y$$

we need to have $x + y = 0$ for a zero-order reaction. This can most obviously be obtained when $x = y = 0$. Then

$$\text{rate} = k[A]^0[B]^0$$

and since any number raised to a zero power is one, the expression for the rate is therefore

$$\text{rate} = k \times 1 \times 1 = k$$

i.e. the rate remains constant throughout the reaction and does not depend on the concentrations of the reactants.

4.3.1   Integration

Integration is essentially the reverse of differentiation, so if we take a function, differentiate it and then integrate it, we should get back to our starting function.

Let us consider what happens if we differentiate the function $f(x) = x^2$. Remember that when we differentiate we multiply by the power and reduce the power by one, so that we obtain the derivative

$$\frac{df(x)}{dx} = 2 \times x^{2-1} = 2x$$

In words, the derivative of x^2 is $2x$. Since integration is the reverse of this, we can also state that:

the integral of $2x$ is x^2

Since the 2 in this expression is simply a constant, we can divide our statement by two so that it becomes:

the integral of x is $\dfrac{x^2}{2}$

Such statements in words are cumbersome, particularly when we have more complicated expressions, and so in mathematical terminology, we write

$$\int x \, dx = \frac{x^2}{2}$$

The precise words to express this are: 'the integral of x with respect to x is equal to x squared divided by 2'.

Using similar arguments, and remembering that

$$\frac{d(x^3)}{dx} = 3x^2$$

we obtain the result

$$\int x^2 \, dx = \frac{x^3}{3}$$

You should begin to see a pattern emerging. In fact, to integrate any polynomial type function, we simply increase the power by one and divide by the new power. This works for positive and negative powers, but not for x^{-1} (alternatively written as $1/x$) since this would involve dividing by zero. In mathematical terminology

$$\int x^n \, dx = \frac{x^{n+1}}{n+1} \quad \text{for } n \neq -1$$

There is one complication in considering integration to be the reverse process of differentiation. Applying the standard methods of differentiation gives us the derivatives

$$\frac{d(x^3 + 1)}{dx} = 3x^2$$

$$\frac{d(x^3 + 2)}{dx} = 3x^2$$

since the derivative of any constant term (1 or 2 in these examples) is zero. If we now try to reverse the process by integrating we obtain

$$\int 3x^2 = \frac{3x^3}{3} = x^3$$

and we obtain no information about the constant term. It is usual when calculating such **indefinite integrals** to include a constant of integration which is usually given the symbol C. We then obtain

$$\int 3x^2 \, dx = x^3 + C$$

C can often be calculated if we know the value of our function $f(x)$ at a certain value of x. These known values are substituted into the equation which can then be solved for C.

More commonly in chemistry, we evaluate what is called the **definite integral**. We are frequently interested in a range of values which we include as limits, and these avoid the need to calculate a constant of integration. For example, if we were asked to calculate the integral of x^3 between $x = 1$ and $x = 2$, we could express this as

$$\int_1^2 x^3 \, dx$$

where the lower and upper limits are included on the integration sign. To integrate, we raise the power of the function x^3 by one and divide by the new power, to obtain $x^4/4$. This expression is enclosed in square brackets with the limits outside and so we write

$$\int_1^2 x^3 \, dx = \left[\frac{x^4}{4} + C \right]_1^2$$

where C is the constant of integration.

We then evaluate the expression at the upper limit and subtract from it the value at the lower limit, and obtain

$$\left[\frac{x^4}{4} + C \right]_1^2 = \left(\frac{2^4}{4} + C \right) - \left(\frac{1^4}{4} + C \right)$$

$$= \left(\frac{16}{4} \right) - \left(\frac{1}{4} \right)$$

$$= \frac{15}{4}$$

this time using rounded brackets, if necessary. Notice that the constants of integration have cancelled and we obtain a definite value for the integral.

Worked example 4.4

The rate equation for a zero-order reaction can be expressed as

$$-\frac{dc}{dt} = k$$

where dc/dt is the rate of change of concentration c of a reactant with time t, and k is the rate constant. Obtain an expression for the concentration c at a time t assuming that the initial concentration value is c_0.

Chemical background

A number of heterogeneous reactions obey zero-order kinetics. These include the decompositions of phosphine or ammonia on hot tungsten, shown in Figure 4.1, and the decomposition of hydrogen iodide on a hot gold wire. This is the case as long as the gas pressure is not too low.

Arsine AsH_3 and stibine SbH_3 both have similar properties to phosphine, but decompose more readily into their elements.

Solution to worked example

As the expression given contains a derivative, we need to reverse the differentiation process by integration. The first step is to separate the variables. This is necessary before we can integrate and involves rearranging the expression so that all terms in the variable c are on one side of the equation and all terms in the variable t are on the other side. In this example, this can be done by multiplying both sides by the term dt to give

$$-dc = k\,dt$$

It is also usual to make the subject (i.e. the left-hand side of the equation) positive, and this can be achieved by multiplying both sides of the equation by -1 to give

$$dc = -k\,dt$$

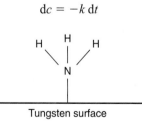

Tungsten surface

Figure 4.1   Ammonia decomposing on the surface of hot tungsten.

We have now separated the variables and so this expression is now in a form which can be integrated. We write

$$\int dc = \int -k\, dt$$

using the integral signs to show integration.

Note that since $-k$ is a constant it can be brought outside the integral sign:

$$\int dc = -k \int dt$$

We now need to assign limits to avoid having to deal with constants of integration. Since the initial concentration is c_0, this will occur when t is zero and so our lower limits are $c = c_0$ and $t = 0$. As we require an expression for c after time t, these are set as our upper limits. This may seem strange at first, but this is a useful tool for obtaining expressions of general use. Including these limits gives us the expression

$$\int_{c_0}^{c} dc = -k \int_{0}^{t} dt$$

It may not be obvious at this stage what quantities we are going to integrate. The terms dc and dt are part of the notation of integration which say 'with respect to c' and 'with respect to t' respectively. Remember however that both the quantities c^0 and t^0 have a value of one, so we can quite justifiably include them in the respective integrals:

$$\int_{c_0}^{c} c^0\, dc = -k \int_{0}^{t} t^0\, dt$$

We can now simply apply the rule that to integrate we increase the power by one and divide by the new power. Thus c^0 integrates to c and t^0 integrates to t. You will find it worthwhile remembering the rule:

$$\int dx = x$$

We now need to evaluate our integrated expression at the specified limits

$$\left[c \right]_{c_0}^{c} = -k \left[t \right]_{0}^{t}$$

We substitute the limits for the quantities in square brackets, and subtract the lower from the upper to obtain

$$c - c_0 = -k(t - 0)$$

$$c - c_0 = -kt$$

Alternatively, this could be multiplied throughout by -1 to give

$$c_0 - c = kt$$

Worked example 4.5

What are the units of the rate constant for a zero-order reaction?

Chemical background

The rate constant has different units for reactions of different order. In contrast, the actual rate of a reaction will always have the units mol dm^{-3} s^{-1} since it is defined as the rate of change of a concentration with time.

The base hydrolysis of nitropentacyanoferrate(III), $Fe(CN)_5NO_2^{3-}$, in aqueous solution follows zero-order kinetics at a pressure of 0.1 MPa and a temperature of 25°C when the concentration of OH is 0.1 mol dm^{-3}, the rate constant being 1.20×10^7 mol dm^{-3} s^{-1}.

Solution to worked example

We need to rearrange the equation

$$c_0 - c = kt$$

to give an expression for k. If we divide both sides of the equation by t, we obtain

$$\frac{c_0 - c}{t} = k$$

The units of k are therefore those of

$$\frac{\text{concentration}}{\text{time}}$$

which are

$$\frac{\text{mol dm}^{-3}}{\text{s}}$$

Since division is the same as multiplying by a reciprocal

$$\frac{1}{\text{s}} = \text{s}^{-1}$$

and so the SI units of the zero-order rate constant are mol dm^{-3} s^{-1}. This is as expected, since for a zero-order reaction the rate of reaction is simply equal to the rate constant.

Worked example 4.6

The half-life of a reaction $t_{1/2}$ is the time taken for the concentration of reactant to fall to half its initial value. What is the half-life for a zero-order reaction?

Chemical background

Although the definition of a half-life is given in terms of the concentration relative to its initial value, it does not actually matter when the starting concentration is taken. The time interval for a particular concentration to be halved will remain constant regardless of the starting time.

Solution to worked example

In this problem, we need to rearrange the integrated rate equation

$$c_0 - c = kt$$

to give an expression for t. This is done by dividing both sides by k to give

$$t = \frac{c_0 - c}{k}$$

From the information given in the question, when t is equal to $t_{1/2}$, c will be equal to $\frac{1}{2}c_0$, i.e. half the initial value. Substituting these values gives us

$$t_{1/2} = \frac{c_0 - \frac{1}{2}c_0}{k}$$

$$= \frac{\frac{1}{2}c_0}{k}$$

$$= \frac{c_0}{2k}$$

Worked example 4.7

If the concentration of a reactant in a zeroth order reaction was monitored as a function of time, what graph would you plot to obtain the value of the rate constant?

Chemical background

Some of the practicalities of monitoring such reactions are illustrated by the catalytic decomposition of hydrogen iodide on the surface of gold, which was reported by Hinshelwood and Prichard in the *Journal of the American Chemical Society* in 1925 (**127**, 1552). There is a difficulty in following the pressure change during this reaction as the hydrogen iodide attacks the mercury in a manometer. This was overcome by using hydrogen gas as a buffer in the manometer, since it has no effect on the rate of reaction.

Solution to worked example

Whenever we are analysing data, we try to plot graphs which give us straight lines. The way in which we can compare an expression with the general straight line equation $y = mx + c$ was outlined in section 2.4.2. Here, x and y are the variables, and m and c are both constants representing the slope of the graph and the intercept on the y axis respectively.

In this case, we need to rearrange the equation obtained in Worked Example 4.4 into this form. Note that c and t are the variables. If we start with

$$c - c_0 = -kt$$

and add c_0 to both sides we get

$$c = -kt + c_0$$

This is now in the required form with c and t as variables and $-k$ and c_0 as constants. If we plot c on the y-axis and t on the x-axis we will obtain a straight line with gradient $-k$ and intercept c_0, as shown in Figure 4.2.

4.4
First-order reactions

The rate equation for a first-order reaction is of the form

$$\text{rate} = k[A]$$

where $[A]$ is the concentration of a reactant and k is the first-order rate constant. The rate of reaction now depends on the concentration of reactant, and as this falls during the course of the reaction, so does the rate.

4.4.1   Integration of $1/x$

We saw in section 4.3.1 that we could integrate any expression of the form x^n as long as $n \neq -1$. The integration rule involves raising the power of x by one and dividing by the new power but, if $n = -1$, you find you need to divide by

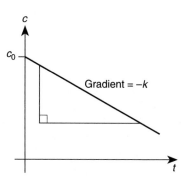

Figure 4.2   Graph of integrated rate equation for a zero-order reaction.

zero and this will not work. In fact, there is another rule for integrating x^{-1} (or $1/x$):

$$\int x^{-1}\,dx = \int \frac{1}{x}\,dx = \int \frac{dx}{x} = \ln x$$

This tells us that when integrating x^{-1} we obtain the natural logarithm, $\ln x$. Notice that the three forms of expressing this integral are all equivalent and you may well meet all of them.

4.4.2   Rules of logarithms

Section 2.4.1 mentioned that, in the past, logarithms were used to assist in the processes of multiplication and division. If we have two numbers X and Y,

$$\log(XY) = \log X + \log Y$$

$$\log\left(\frac{X}{Y}\right) = \log X - \log Y$$

These rules apply regardless of the base to which the logarithms are taken. They replace the multiplication process of X and Y by addition of their logs, and the division process of X and Y by subtraction of their logs, if the logarithms of the required numbers can be obtained.

Notice that if $X = Y$ we have

$$\log XX = \log X + \log X = 2 \log X$$

so

$$\log X^2 = 2 \log X$$

Also

$$\log \frac{X}{X} = \log X - \log X = 0$$

so

$$\log 1 = 0$$

Generally, $\log X^n = n \log X$ for all values of n.

Worked example 4.8

The rate equation for a first-order reaction can be expressed as

$$-\frac{dc}{dt} = kc$$

where dc/dt is the rate of change of concentration c of a reactant with time t, and k is the rate constant. Obtain an expression for the concentration c at a time t assuming that the initial concentration value is c_0.

Chemical background

Many reactions obey first-order kinetics. Examples include

$$2N_2O_{5(g)} \rightarrow 4NO_{2(g)} + O_{2(g)}$$

and

$$SO_2Cl_2 \rightleftharpoons SO_2 + Cl_2$$

Sulphuryl chloride, which is a colourless fuming liquid, hydrolyses vigorously to form sulphuric and hydrochloric acid. It is used as a chlorinating agent.

Solution to worked example

Most of the stages in setting up this problem are identical to those in the case of the zero-order reaction in Worked Example 4.4. To integrate we must first separate the variables. If the terms containing variable c are to appear on the left-hand side as before it is necessary to divide both sides by c and multiply both sides by dt before adding the integrals in symbols complete with initial and final values of c and t. This then results in the expression

$$\int_{c_0}^{c} \frac{dc}{c} = -k \int_{0}^{t} dt$$

As before,

$$\int dt = t$$

while we now know that

$$\int \frac{1}{c} dc = \ln c$$

We then obtain

$$\left[\ln c\right]_{c_0}^{c} = -k\left[t\right]_{0}^{t}$$

which on substituting the upper and lower limits into the square brackets gives

$$\ln c - \ln c_0 = -k(t - 0)$$

$$= -kt$$

We saw earlier in this section that

$$\log X - \log Y = \log\left(\frac{X}{Y}\right)$$

so, in this case

$$\ln c - \ln c_0 = \ln\left(\frac{c}{c_0}\right)$$

Putting this into the integrated first-order rate equation gives us the result

$$\ln\left(\frac{c}{c_0}\right) = -kt$$

Worked example 4.9

What are the units of the first-order rate constant?

Chemical background

The complex $[Cr(en)_3]^{3+}$ has been observed in three distinct conformations in different crystal structures.

An example of a first-order rate constant is the value of $3.3 \times 10^{-4}\,\mathrm{s}^{-1}$ for the *cis-trans* isomerization of $Cr(en)_2(OH)_2^+$. The symbol en stands for the ligand ethylenediamine, $NH_2CH_2CH_2NH_2$.

Solution to worked example

As in the case of Worked Example 4.5 for a zero-order reaction, we need to rearrange our integrated rate equation to make the rate constant, k, the subject. Dividing both sides of the equation

$$\ln\left(\frac{c}{c_0}\right) = -kt$$

by $-t$, gives

$$-\frac{\ln\left(\frac{c}{c_0}\right)}{t} = k$$

Since logarithmic quantities are dimensionless, i.e. they have no units, k will have units equivalent to those of $1/t$ or t^{-1}, i.e. s^{-1}.

Worked example 4.10

What is the half-life of a first-order reaction?

Chemical background

The fact that different expressions emerge for the half-lives of reactions of different orders leads to a very quick way of determining whether a reaction is first order or not. The half-lives of zero- and second-order reactions depend on the initial reactant concentration, whereas the half-life of the first-order reaction does not.

Solution to worked example

We saw in Worked Example 4.6 that the half-life $t_{1/2}$ of a reaction is the time taken for the concentration to fall to half its initial value. Rearranging the integrated rate equation for a first-order reaction to give an expression for t gives

$$t = -\frac{1}{k}\ln\left(\frac{c}{c_0}\right)$$

Now $\log X^n = n \log X$ so when $n = -1$,

$$\log X^{-1} = -\log X$$

so

$$t = \frac{1}{k}\ln\left(\frac{c}{c_0}\right)^{-1}$$

$$= \frac{1}{k}\ln\left(\frac{c_0}{c}\right)$$

Sustituting $t = t_{1/2}$ and $c = \frac{1}{2}c_0$ now gives

$$t_{1/2} = \frac{1}{k}\ln\left(\frac{c_0}{\frac{1}{2}c_0}\right)$$

$$= \frac{\ln 2}{k}$$

since the c_0s cancel and $1 \div \frac{1}{2} = 2$.

Worked example 4.11

Solid N_2O_5 consists of NO_2^+ and NO_3^- ions. It is colourless and stable below 0°C. At higher temperatures, it decomposes slowly into N_2O_4 and O_2.

A determination of the pressure of gaseous N_2O_5 as a function of time gave the values

t/s	0	1200	4800	7200
$\dfrac{p}{\text{mm Hg}}$	350	190	34	10

If the decomposition of N_2O_5 is a first-order reaction, what is the value of the rate constant?

Chemical background

This is a frequently quoted example of a first-order reaction, the equation for the decomposition being

$$2N_2O_{5(g)} \rightarrow 4NO_{2(g)} + O_{2(g)}$$

The total pressure can be measured by connecting the flask to a manometer, but it is then necessary to calculate the contribution from N_2O_5.

Pressure is a useful quantity to determine since it is proportional to concentration. If we consider the ideal gas equation

$$pV = nRT$$

and divide both sides by V, we obtain

$$p = \left(\frac{n}{V}\right)RT$$

The quantity n/V has the units of mol per unit volume, so is actually a measure of concentration c. We can therefore write

$$p = cRT$$

and since R is a constant at a constant temperature p will be proportional to c.

Solution to worked example

Since this is a first-order reaction, we will use the integrated rate equation

$$\ln c - \ln c_0 = -kt$$

If the concentration c is proportional to pressure p, we can write

$$c = Kp$$

and

$$c_0 = Kp_0$$

where K (in this case equal to $1/(RT)$) is the constant of porportionality as explained in section 2.5.3 and p_0 is the initial pressure. Substituting into the integrated rate equation then gives

$$\ln(Kp) - \ln(Kp_0) = -kt$$

We saw earlier that

$$\ln(XY) = \ln X + \ln Y$$

and so

$$\ln(Kp) = \ln K + \ln p$$

and

$$\ln(Kp_0) = \ln K + \ln p_0$$

can be substituted into the left-hand side of the equation

$$(\ln K + \ln p) - (\ln K + \ln p_0) = -kt$$

The brackets can be removed, ensuring that the minus sign is applied to both terms in the second bracket, to give

$$\ln K + \ln p - \ln K - \ln p_0 = -kt$$

The terms $\ln K$ and $-\ln K$ cancel, to leave

$$\ln p - \ln p_0 = -kt$$

Adding $\ln p_0$ to both sides of the equation gives

$$\ln p = -kt + \ln p_0$$

Since p_0 is a constant $\ln p_0$ is also a constant, and so we now have an equation which is of the form $y = mx + c$, as discussed in section 2.4.2. Comparing corresponding quantities in the two forms of the equation shows that if $\ln p$ is plotted on the y-axis and t on the x-axis, we will obtain a straight line graph with gradient $-k$.

The mechanics of graph plotting were outlined in section 2.4.3. We first need to draw up a table containing values of $\ln p$ and t. We will actually calculate the values of $\ln(p/\text{mm Hg})$, since we can only take the logarithm of a dimensionless quantity which has no units. This actually has the effect of introducing a further constant, but does not affect the value of the gradient.

The first value in the table has

$$p/\text{mm Hg} = 350$$

so if we use a calculator we obtain

$$\ln(p/\text{mm Hg}) = \ln 350 \quad \text{▦}$$

$$\approx 5.86$$

We can obtain other values in a similar fashion, and the data we need to plot can be tabulated as

t/s	0	1200	4800	7200
$\ln\left(\dfrac{p}{mm\ Hg}\right)$	5.86	5.25	3.53	2.30

The graph of $\ln(p/mm\ Hg)$ against t is shown in Figure 4.3, from which we can obtain the gradient.

$$\text{gradient} = \frac{\text{increase in } \ln(p/mm\ Hg)}{\text{increase in } t}$$

$$= \frac{2.00 - 5.85}{(7900 - 0)\ s}$$

$$= \frac{-3.85}{7900\ s}$$

$$\simeq -4.9 \times 10^{-4}\ s^{-1}$$

Now, we have already seen that the gradient is equal to $-k$, so

$$k = 4.9 \times 10^{-4}\ s^{-1}$$

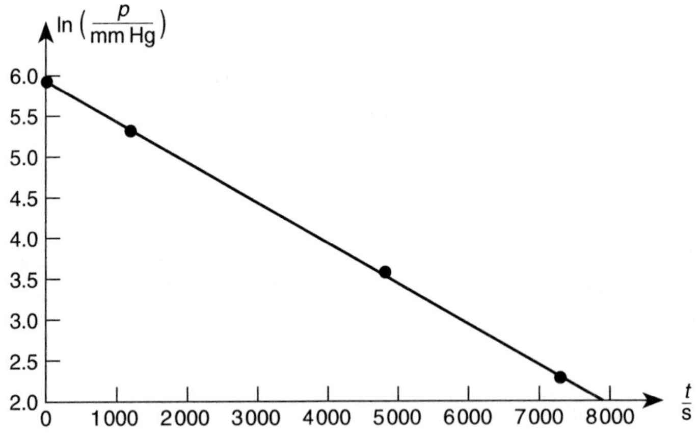

Figure 4.3   Plot of $\ln(p/mm\ Hg)$ against time for the decomposition of N_2O_5.

4.5
Second-order reactions   If we consider the generalized expression for the rate of a reaction

$$\text{rate} = k[A]^x[B]^y$$

there are two ways in which it is possible to achieve an overall order of 2, i.e. to satisfy the equation

$$x + y = 2$$

Case 1: x = 2, y = 0

If $x = 2$ and $y = 0$ (or $x = 0$, and $y = 2$), the rate equation is

$$\text{rate} = k[A]^2$$

This would arise in a reaction of the type

$$A + B \rightarrow \text{products}$$

in which the starting concentrations of A and B were equal.

Case 2: x = y = 1

If $x = y = 1$, the equation is

$$\text{rate} = k[A][B]$$

The first of these two cases is rather simpler to deal with than the second.

Worked example 4.12

The rate equation for a second-order reaction in which the initial concentrations of the reactants are equal can be expressed as

$$-\frac{dc}{dt} = kc^2$$

where dc/dt is the rate of change of concentration c of a reactant with time t, and k is the rate constant. Obtain an expression for the concentration c at a time t, assuming that the initial concentration value is c_0.

Chemical background

Typical reactions which obey second-order kinetics are

$$H_{2(g)} + I_{2(g)} \rightarrow 2HI_{(g)}$$

and

$$CH_3Br + OH^- \rightarrow CH_3OH + Br^-$$

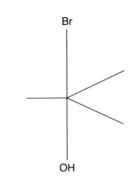

Figure 4.4   Transition state in the S_N2 reaction.

The second of these is an example of an S_N2 (substitution nucleophilic bimolecular) reaction, which involves a transition state in which carbon is partially bonded to both $-OH$ and $-Br$, as shown in Figure 4.4.

Solution to worked example

The initial stages involve rearranging the given equation to separate the terms, i.e. to have all terms in c on the left-hand side and all those in t on the right-hand side. Then we can integrate and assign limits as in Worked Examples 4.4 and 4.8. This results in us being required to solve the equation

$$\int_{c_0}^{c} \frac{dc}{c^2} = -k \int_{0}^{t} dt$$

The integration on the right-hand side is straightforward and identical to the previous problems.

$$\int dt = t + \text{constant}$$

That of $1/c^2$ requires a little more thought, but is not difficult once we realize that $1/c^2$ can be written as c^{-2}. Raising the power by one gives us c^{-1} (not c^{-3} which would be a reduction of the power by one) so we need to divide c^{-1} by -1. Note that dividing by -1, is the same as multiplying by -1 because

$$x \div -1 = \frac{x}{-1} = -x \quad \text{and} \quad x \times -1 = -x$$

Consequently, we need to evaluate the expression

$$\left[-\frac{1}{c} \right]_{c_0}^{c} = -k \left[t \right]_{0}^{t}$$

If we substitute the upper limits and then substract the value obtained with the lower limits, we get

$$-\frac{1}{c} - \left(-\frac{1}{c_0}\right) = -k(t - 0)$$

which, if we multiply every term by -1, becomes

$$\frac{1}{c} - \frac{1}{c_0} = kt$$

Worked example 4.13

What are the units of the second-order rate constant?

Chemical background

An example of the second-order rate constant is the value of 2.42×10^{-2} dm^3 mol^{-1} s^{-1} for the reaction

$$H_2 + I_2 \rightarrow 2HI$$

An alternative method for synthesizing hydrogen iodide is by reducing iodine with hydrogen sulphide. Hydrogen compounds generally become less stable as a group of the periodic table is descended, so HF will be more stable than HI.

Solution to worked example

As with other integrated rate equations, we need to rearrange to make the rate constant k the subject of the equation. We can do this by dividing both sides of the equation by t, to give

$$k = \frac{1}{t}\left(\frac{1}{c} - \frac{1}{c_0}\right)$$

The units we would get from substituting into this equation would therefore be

$$\frac{1}{s}\left(\frac{1}{\text{mol dm}^{-3}}\right)$$

Note that

$$\frac{1}{s} = s^{-1} \quad \text{and} \quad \frac{1}{\text{dm}^{-3}} = \text{dm}^3$$

Then the units can be expressed as dm^3 mol^{-1} s^{-1}. We write the units with the positive powers before those with the negative powers because, when expressed verbally we replace negative powers by the word 'per'. Thus dm^3 mol^{-1} s^{-1} would be said: dm cubed per mole per second.

Worked example 4.14

If the concentration of a reactant in a second-order reaction, with equal concentrations of reactants, was monitored as a function of time, what graph would you plot to obtain the value of the rate constant?

Chemical background

The reactions of the carbonyl compounds acetaldehyde, propionaldehyde and acetone with HCN to give cyanohydrins were found to be second order in a study reported in the *Journal of the American Chemical Society* in 1953 (**75**, 3106). Measurements were performed in acetate buffers under various conditions to ascertain the influence of ionic strength, pH and buffer composition on the rate constants.

Solution to worked example

In order to obtain a value of the rate constant k, we need to plot the data in such a way that we obtain a straight line. This can be done by rearranging the integrated rate equation above into the form of the equation of a straight line graph, $y = mx + c$, as outlined in section 2.4.2.
    Starting with the equation

$$\frac{1}{c} - \frac{1}{c_0} = kt$$

we can add the constant term $1/c_0$ to both sides to give

$$\frac{1}{c} = kt + \frac{1}{c_0}$$

This is now in the form we want, with $1/c$ corresponding to y and t corresponding to x. If a graph of $1/c$ against t is plotted, we will obtain a straight line with gradient k and intercept $1/c_0$ as shown in Figure 4.5.

Worked example 4.15

Methyl acetate is an ester which can be formed by the reaction of acetic anhydride with methanol. It melts at $-98°C$ and boils at $57.5°C$.

During the hydrolysis of methyl acetate, aliquots of the reaction mixture were withdrawn at intervals and titrated with sodium hydroxide, to give the following results:

t/min	5.64	20.69	45.74	75.76
V/cm^3	26.52	27.98	29.88	31.99

where V is the titre of sodium hydroxide. Determine the order of this reaction.

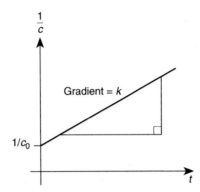

Figure 4.5   Graph of integrated rate equation for a second-order reaction.

Chemical background

The data given refer to the acid hydrolysis of methyl acetate, which can be represented by the equation

$$CH_3COOCH_3 + H_2O \rightleftharpoons CH_3COOH + CH_3OH$$

The carbonyl oxygen becomes protonated and so the carbonyl carbon atom is more susceptible to nucleophilic attack by the water molecule. There are likely to be several tetrahedral intermediates in this reaction.

Solution to worked example

We have seen earlier that there are three distinct plots which will give a straight line in reactions we are likely to meet. To summarize, we need to plot

- c against t for a zero-order reaction
- $\ln c$ against t for a first-order reaction
- $1/c$ against t for a second-order reaction

To determine the order of a reaction such as this, we need to perform the three plots and identify which one gives the straight line. Hopefully, this will be easy to spot.

In this example, there are further complications. We are not given the concentration of methyl acetate, but the titre of sodium hydroxide. However, since the rate of increase of methanoic acid is numerically equivalent to the rate of decrease of methyl acetate, these quantities are proportional and we can write

$$V = Kc$$

where K is the constant of proportionality and so

$$V_0 = Kc_0$$

If we now substitute into the integrated rate equation for a zero-order reaction

$$c = -kt + c_0$$

we have

$$\frac{V}{K} = -kt + \frac{V_0}{K}$$

We can multiply both sides by the proportionality constant K to give

$$V = -kKt + V_0$$

so in fact a plot of V against t will be a straight line having gradient kK and intercept V_0.

In a similar way, it is possible to show that straight line plots will be obtained for a reaction of appropriate order if a quantity which is proportional to the concentration is used instead of the concentration itself. In this problem, it is therefore possible to use the titre values directly, rather than converting them to concentrations.

A further complication is that we do not have the values of V_0, the titre value at zero time. However, as the time values are all relative, we can subtract the first value of 5.64 from each, which is equivalent to assuming that the reaction began 5.64 minutes later. If we do this, the data we need to use become

t/min	0.00	15.05	40.10	70.12
V/cm^3	26.52	27.98	29.88	31.99

To consider the first- and second-order plots, we need the values of

$$\ln\!\left(\frac{V}{cm^3}\right) \quad \text{and} \quad \frac{1}{V}$$

If we consider the first point, we have

$$V = 26.52 \text{ cm}^3$$

$$\frac{V}{cm^3} = 26.52$$

$$\ln\!\left(\frac{V}{cm^3}\right) \simeq 3.278$$

$$\frac{V}{cm^3} = 26.52$$

$$\frac{1}{V} \simeq 0.03771 \text{ cm}^{-3}$$

The full table with these quantities is as follows:

t/min	0.00	15.05	40.10	70.12
V/cm^3	26.52	27.98	29.88	31.99
$\ln(V/\text{cm}^3)$	3.278	3.331	3.397	3.465
V^{-1}/cm^{-3}	0.03771	0.03574	0.03346	0.03126

The graphs of V against t, $\ln V$ against t and $1/V$ against t are shown in Figure 4.6. Since only the plot of $\ln V$ against t gives a straight line, the reaction must be first order.

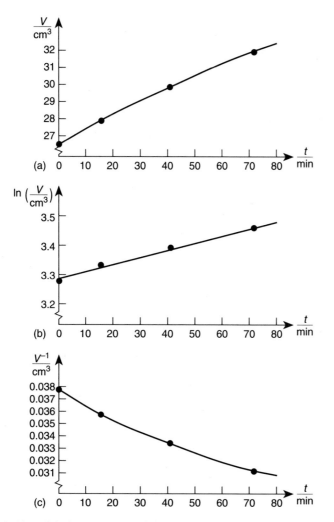

Figure 4.6 Plots of the integrated rate equations for the hydrolysis of methyl acetate: (a) zero-order; (b) first-order; and (c) second-order.

To analyse second-order kinetics in the situation when the starting concentrations of the two reactants are not equal, we need some additional mathematical techniques.

4.5.1 *Partial fractions*

Partial fractions are useful for expressing a fractional product as the sum of two fractions, which in some situations may be easier to deal with. For example, if we have an expression such as

$$\frac{1}{(x + 1)(x + 2)}$$

it may be preferable to express it as the sum of two fractions such as

$$\frac{A}{x + 1} + \frac{B}{x + 2}$$

where A and B are constants. Determining the partial fractions, in this case, is simply a matter of finding values for the constants A and B. Since these expressions are identical, we can write the **identity**

$$\frac{1}{(x + 1)(x + 2)} \equiv \frac{A}{(x + 1)} + \frac{B}{(x + 2)}$$

Notice the use of the $\equiv$ symbol here. It is used instead of $=$ because this expression is true for *all* values of x. An equation, on the other hand, is only true for certain values of x.

If we now consider the right-hand side of this expression, we can multiply the top and bottom of the first term by $(x + 2)$ and the top and bottom of the second term by $(x + 1)$:

$$\frac{1}{(x + 1)(x + 2)} \equiv \frac{A(x + 2)}{(x + 1)(x + 2)} + \frac{B(x + 1)}{(x + 2)(x + 1)}$$

because multiplying both the top and bottom of a fraction by the same quantity does not change its value. The two terms on the right-hand side now have the same denominators so we can now write

$$\frac{1}{(x + 1)(x + 2)} \equiv \frac{A(x + 2) + B(x + 1)}{(x + 1)(x + 2)}$$

Multiplying both expressions by $(x + 1)(x + 2)$, gives

$$1 \equiv A(x + 2) + B(x + 1)$$

Now since this is an identity, it will be true for any value of x. If we choose to set $x = -1$, the bracket containing $x + 1$ will become zero and we have

$$A(-1 + 2) \equiv 1$$

$$A \equiv 1$$

Similarly if we set $x = -2$, the bracket containing $x + 2$ becomes zero and we have

$$B(-2 + 1) \equiv 1$$

$$-B \equiv 1$$

$$B \equiv -1$$

Substituting into our general expression for the partial fractions then gives

$$\frac{1}{(x + 1)(x + 2)} \equiv \frac{1}{x + 1} + \frac{-1}{x + 2}$$

$$\equiv \frac{1}{x + 1} - \frac{1}{x + 2}$$

There are more complicated cases of partial fractions, but this knowledge is sufficient for our present purposes.

4.5.2 Differentiation of logarithmic functions and integration of fractions

We have seen previously that

$$\frac{d(\ln x)}{dx} = \frac{1}{x}$$

and that

$$\int \frac{dx}{x} = \ln x + C$$

where C is the constant of integration.

Suppose, however, that we wished to differentiate $\ln(3 - x)$, i.e. to calculate

$$\frac{d(\ln(3 - x))}{dx}$$

You might guess that because we are differentiating a ln function this will give us the function

$$\frac{1}{3 - x}$$

This is partly true, but we need also to differentiate the quantity in the brackets.

$$\frac{d(3 - x)}{dx} = \frac{d(3)}{dx} + \frac{d(-x)}{dx}$$

$$= 0 + -1$$

$$= -1$$

and the whole derivative is

$$-1 \times \frac{1}{3 - x}$$

which can be multiplied top and bottom by -1 to give

$$\frac{1}{x - 3}$$

We call

$$f(x) = \ln(3 - x)$$

a **function of a function**: the logarithmic function of $3 - x$. Replacing 3 by a general constant a, we obtain the rule

$$\frac{d(\ln(a - x))}{dx} = \frac{1}{x - a}$$

Since integration is the reverse of differentiation, we also have

$$\int \frac{dx}{x - a} = \ln(a - x) + C$$

where C is the constant of integration.

There is a complication in that we can only take the logarithm of a positive number, and the correct expression for dealing with general integrations of this type is

$$\int \frac{dx}{ax + b} = \frac{1}{a} \ln|ax + b| + C$$

The vertical bars simply denote that we take the modulus of the quantity enclosed between them. This is the value taking no account of its sign, as we saw previously in section 3.1.1. In the examples we meet in chemistry the modulus symbols are frequently omitted, so we would write

$$\int \frac{dx}{ax + b} = \frac{1}{a} \ln(ax + b) + C$$

Worked example 4.16

The rate equation for a second-order reaction in which the initial concentrations of the reactants are different can be expressed as

$$-\frac{da}{dt} = kab$$

where da/dt is the rate of change of concentration a of one reactant with time t, b is the concentration of the second reactant, and k is the rate constant. Obtain an expression for the concentrations a and b at a time t assuming that the initial concentration values are a_0 and b_0.

Chemical background

The analysis of general second-order reactions of this type is quite complicated. It is also possible to have reactions of higher order, but the analysis of these will not be considered here as they are rarely encountered. One example of a reaction exhibiting third-order kinetics is that between nitric oxide and chlorine

$$2NO + Cl_2 \rightarrow 2NOCl$$

which obeys the rate equation

$$-\frac{d[NO]}{dt} = k[NO]^2[Cl_2]$$

Bacterial action is responsible for 80% of NO in the atmosphere, while the rest is produced by combustion of various types. Both NO and NO_2 are toxic and take part in photochemical reactions in the atmosphere – they are known as NO_x gases.

Solution to worked example

As previously, we begin by separating the variables, i.e. rearranging the expression given so that all the concentration terms are on the left, with the rate constant and the time on the right. This gives us the expression

$$\int \frac{da}{ab} = -k \int dt$$

where the rate constant k is taken outside the integral on the right-hand side.
   We immediately have a problem, since we need to integrate

$$\frac{1}{ab}$$

with respect to a and this cannot be done as the expression stands because it also contains the variable b. To overcome this difficulty, remember that the generalized equation for this reaction is

$$A + B \rightarrow products$$

The initial concentrations of A and B are given as a_0 and b_0 respectively. Suppose that after time t the concentration of A has reduced by x; if A and B react in an equimolar ratio then the concentration of B will also have reduced by x and we can write for the concentrations of A and B

$$a = a_0 - x$$
$$b = b_0 - x$$

Since a_0 and b_0 are constants, we have now expressed the concentrations of A and B in terms of a single common variable x. If we differentiate the first equation, since a_0 is a constant then

$$\frac{da}{dx} = -1$$

and so we can replace da by $-dx$, in the integration.

We are now able to substitute for a, b and da to obtain

$$\int \frac{dx}{(a_0 - x)(b_0 - x)} = k \int dt$$

noticing that the minus signs cancel on the right-hand side. This expression now needs to be integrated between appropriate limits. Clearly when $t = 0$ none of A or B has reacted so $x = 0$. We want to obtain an expression giving x after time t so we use these values as our upper limits:

$$\int_0^x \frac{dx}{(a_0 - x)(b_0 - x)} = k \int_0^t dt$$

Integrating the right-hand side is easy.

$$\int_0^t dt = \left[t \right]_0^t = t - 0 = t$$

However, we now have the problem of integrating the expression on the left, which is not so easy. We saw earlier how an expression of this type could be written in terms of two partial fractions. As before, we start by setting up an identity in terms of two unknown constants A and B:

$$\frac{1}{(a_0 - x)(b_0 - x)} \equiv \frac{A}{a_0 - x} + \frac{B}{b_0 - x}$$

We multiply the first term, top and bottom, by $(b_0 - x)$ and the second term, top and bottom, by $(a_0 - x)$; neither of these operations changes the value of the expression but it does produce fractions with the same denominators.

$$\frac{1}{(a_0 - x)(b_0 - x)} \equiv \frac{A(b_0 - x)}{(a_0 - x)(b_0 - x)} + \frac{B(a_0 - x)}{(b_0 - x)(a_0 - x)}$$

Since the quantity on the bottom is the same for every term in this expression we multiply all terms by $(a_0 - x)(b_0 - x)$ to give

$$1 \equiv A(b_0 - x) + B(a_0 - x)$$

This expression is true for all values of x, so we set $x = b_0$ to make the first bracket zero and to leave

$$1 \equiv B(a_0 - b_0)$$

This rearranges to give

$$B \equiv \frac{1}{a_0 - b_0}$$

Similarly, setting $x = a_0$ makes the second bracket zero and we have

$$1 \equiv A(b_0 - a_0)$$

which rearranges to

$$A \equiv \frac{1}{b_0 - a_0}$$

It is worth noticing that $A \equiv -B$, as this will allow some simplification later. Substituting in our original identity for the partial fractions gives

$$\frac{1}{(a_0 - x)(b_0 - x)} \equiv \frac{1}{(b_0 - a_0)(a_0 - x)} + \frac{1}{(a_0 - b_0)(b_0 - x)}$$

If we multiply the top and bottom of the second term on the right by -1 this becomes

$$\frac{1}{(a_0 - x)(b_0 - x)} \equiv \frac{1}{(b_0 - a_0)(a_0 - x)} - \frac{1}{(b_0 - a_0)(b_0 - x)}$$

We can then introduce brackets on the right-hand side and take the common factor of $(b_0 - a_0)$ outside the new brackets to give

$$\frac{1}{(a_0 - x)(b_0 - x)} = \frac{1}{(b_0 - a_0)}\left(\frac{1}{(a_0 - x)} - \frac{1}{(b_0 - x)}\right)$$

Bearing in mind that $b_0 - a_0$, and consequently $1/(b_0 - a_0)$, is a constant, we can now write our equation to be integrated as

$$\frac{1}{b_0 - a_0}\int_0^x \left(\frac{1}{(a_0 - x)} - \frac{1}{(b_0 - x)}\right) dx = k\int_0^t dt$$

or

$$\frac{1}{b_0 - a_0}\int_0^x \frac{dx}{a_0 - x} - \frac{1}{b_0 - a_0}\int \frac{dx}{b_0 - x} = k\int_0^t dt$$

We saw earlier that

$$\int \frac{dx}{x - a} = \ln(x - a) + C$$

and it follows that if we multiply through by -1 we obtain

$$\int \frac{dx}{a-x} = -\ln(a-x) + C'$$

where C' is, of course, a different constant of integration. We can now calculate the integrals required for solving the second-order rate equation:

$$\int_0^x \frac{dx}{a_0-x} = -\left[\ln(a_0-x)\right]_0^x$$

$$= -\ln(a_0-x) - (-\ln a_0)$$

$$= -\ln(a_0-x) + \ln a_0$$

$$= \ln a_0 - \ln(a_0-x)$$

$$= \ln\left(\frac{a_0}{a_0-x}\right)$$

since as we saw in section 4.4.2

$$\ln D - \ln E = \ln\left(\frac{D}{E}\right)$$

Similarly, we obtain

$$\int_0^x \frac{dx}{b_0-x} = \ln\left(\frac{b_0}{b_0-x}\right)$$

and finally we have

$$\int_0^t dt = t$$

as we have seen for the other rate equations. Putting these into the equation leads to

$$\frac{1}{b_0-a_0}\left(\ln\left(\frac{a_0}{a_0-x}\right) - \ln\left(\frac{b_0}{b_0-x}\right)\right) = kt$$

where we can again write the difference of the two logarithmic terms as a logarithm of a quotient

$$\frac{1}{b_0-a_0}\left(\ln\left(\frac{a_0}{a_0-x}\right)\left(\frac{b_0}{b_0-x}\right)^{-1}\right) = kt$$

and since x^{-1} is equal to $1/x$ this can be rewritten as

$$\frac{1}{b_0-a_0}\left(\ln\frac{a_0(b_0-x)}{b_0(a_0-x)}\right) = kt$$

Remembering that we initially set the concentrations of A and B after time t to be

$$a = a_0 - x \quad \text{and} \quad b = b_0 - x$$

this can be written as

$$\frac{1}{b_0 - a_0} \ln\left(\frac{a_0 b}{a b_0}\right) = kt$$

Notice also that since $\ln x^{-1} = -\ln x$, we can multiply the top and bottom of the left-hand side of this equation by -1 to give

$$\frac{1}{a_0 - b_0} \ln\left(\frac{a_0 b}{a b_0}\right)^{-1} = kt$$

which then becomes

$$\frac{1}{a_0 - b_0} \ln\left(\frac{a b_0}{a_0 b}\right) = kt$$

4.6
The Arrhenius equation

So far we have considered the rate constant k to be constant under specified conditions. This is in fact true as long as the temperature remains constant but, as you might expect, the rate of a reaction increases with the temperature. The relationship between rate constant and temperature is described by an exponential function.

In fact, such a law can also be used to describe other processes, one of these being the chirping of crickets. This may seem surprising at first, until we realise that the chirping is itself controlled by a chemical reaction.

4.6.1   The exponential function

The exponential function can be defined in terms of our usual notation as

$$f(x) = e^x$$

where e has the value 2.7183 to four decimal places. This value is not exact, but we rarely need to worry about the numerical value of e. When complicated expressions are involved, you will also see the exponential function written as

$$f(x) = \exp(x)$$

This has precisely the same meaning as the previous equation and is simply a useful device for improving the clarity of some mathematical expressions.

Scientific calculators allow easy calculation of exponential functions.

It is worth noting that it is only possible to take the exponential of a dimensionless number which has no units. This should always be checked before attempting to calculate an exponential term in a calculation.

4.6.2   Inverse functions

The inverse of a function $f(x)$ is generally known as arc $f(x)$ and simply reverses the effect of the original function. For example, if a function $f(x)$ is defined by

$$f(x) = x + 10$$

this means 'take a value of x and add 10 to it'. To reverse this we must therefore subtract 10 and we write

$$\text{arc } f(x) = x - 10$$

Notice that

$$\text{arc } f(f(x)) = f(x) - 10$$
$$= x + 10 - 10$$
$$= x$$

which is a general result for a function and its inverse.

The natural logarithmic and exponential functions are actually the inverses of each other, so that

$$\text{arc } \ln(x) = e^x \quad \text{and} \quad \text{arc } e^x = \ln x$$

It also follows that

$$e^{\ln x} = x \quad \text{and} \quad \ln e^x = x$$

Worked example 4.17

NO$_2$ can be formed in the laboratory by the catalytic oxidation of ammonia gas by oxygen from the air, followed by the exothermic oxidation of nitric oxide to nitrogen dioxide.

The value of the rate constant k for the decomposition of nitrogen dioxide was found to vary with absolute temperature T according to the Arrhenius equation

$$k = A \exp\left(\frac{-E_a}{RT}\right)$$

where A is a constant called the pre-exponential factor, E_a is the activation energy and R is the ideal gas constant which can be taken as $8.31 \text{ J K}^{-1} \text{ mol}^{-1}$. Use the data below to determine the value of the activation energy E_a:

$\dfrac{T}{K}$	593	604	628	652	657
$\dfrac{k}{(\text{dm}^3 \text{ mol}^{-1} \text{ s}^{-1})}$	0.523	0.751	1.70	4.01	5.01

Chemical background

For some reactions involving free atoms or radicals, there is a very small activation energy and a more accurate treatment of the temperature dependence must be used. This frequently involves using an equation of the form

$$k = AT^n \exp\left(-\frac{E}{RT}\right)$$

where the value of n depends on the nature of the reaction and the theory being used in the analysis. The activation energy E in this equation is related to that in the Arrhenius equation, E_a, by the equation

$$E = E_a - nRT$$

Solution to worked example

A straightforward plot of k against T is shown in Figure 4.7(a) and is obviously curved. As usual, we would like to obtain a linear plot, and this can be done by taking the natural logarithm of each side of the Arrhenius equation. This gives us

$$\ln k = \ln\left(A \exp\left(\frac{-E_a}{RT}\right)\right)$$

The right-hand side of this equation now has the logarithm of a product, and since

$$\ln(XY) = \ln X + \ln Y$$

we can rewrite the equation as

$$\ln k = \ln A + \ln\left(\exp\left(\frac{-E_a}{RT}\right)\right)$$

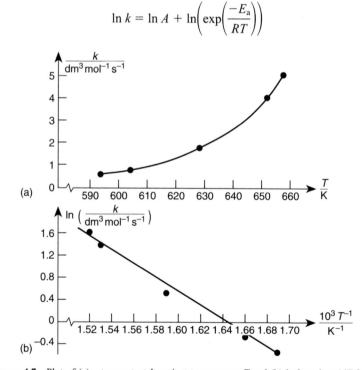

Figure 4.7 Plot of (a) rate constant k against temperature T and (b) $\ln k$ against $1/T$ for the decomposition of nitrogen dioxide.

We have also seen that $\ln e^x = x$ and so our modified Arrhenius equation becomes

$$\ln k = \ln A - \frac{E_a}{RT}$$

Reversing the order of the terms on the right, and grouping the constants E_a and R in brackets gives us

$$\ln k = -\left(\frac{E_a}{R}\right)\left(\frac{1}{T}\right) + \ln A$$

which is equivalent to the general equation of a straight line $y = mx + c$ with $\ln k$ being equivalent to y, $-E_a/R$ being equivalent to the gradient m, $1/T$ being equivalent to x, and $\ln A$ being equivalent to the intercept c. We therefore need to plot $\ln k$ against $1/T$ to obtain a straight line with gradient $-E_a/R$.

Since we can only take the logarithm of a dimensionless number we will actually calculate values of

$$\ln\left(\frac{k}{dm^3 \, mol^{-1} \, s^{-1}}\right)$$

For the first point,

$$k = 0.523 \, dm^3 \, mol^{-1} \, s^{-1}$$

so

$$\frac{k}{dm^3 \, mol^{-1} \, s^{-1}} = 0.523$$

and using a calculator gives

$$\ln\left(\frac{k}{dm^3 \, mol^{-1} \, s^{-1}}\right) \simeq -0.648$$

The temperature values are given in units of K so directly give the values of T. The first point has $T = 593$ K and the calculator gives

$$T^{-1} = \frac{1}{T} \simeq 1.69 \times 10^{-3} \, K^{-1}$$

It is worth looking at how this expression can be rearranged to assist in the tabulation of data. It is neater to incorporate the power of ten into the table heading, rather than including it for every value. The remaining values of T^{-1} are calculated in the same way and can be tabulated as follows.

$\ln(k/dm^3 \, mol^{-1} \, s^{-1})$	-0.648	-0.286	0.531	1.39	1.61
$10^3 \, T^{-1}/K^{-1}$	1.69	1.66	1.59	1.53	1.52

The plot of $\ln(k/\mathrm{dm^3\,mol^{-1}\,s^{-1}})$ against T^{-1} is shown in Figure 4.7(b). From this, we can calculate

$$\mathrm{gradient} = \frac{\text{increase in } \ln k}{\text{increase in } T^{-1}}$$

$$= \frac{-0.65 - 1.85}{(1.70 - 1.50) \times 10^{-3}\,\mathrm{K^{-1}}}$$

$$= \frac{-2.50}{0.20 \times 10^{-3}\,\mathrm{K^{-1}}}$$

$$= -1.25 \times 10^4\,\mathrm{K}$$

We saw earlier that the gradient is equal to $-E_a/R$ so that

$$-\frac{E_a}{R} = -1.25 \times 10^4\,\mathrm{K}$$

The negative signs on either side of the equation cancel, and multiplying both sides by the ideal gas constant R gives

$$E_a = 1.25 \times 10^4\,\mathrm{K} \times R$$

Substituting the value of R gives

$$E_a = 1.25 \times 10^4\,\mathrm{K} \times 8.31\,\mathrm{J\,K^{-1}\,mol^{-1}}$$

$$\simeq 1.04 \times 10^5\,\mathrm{J\,mol^{-1}}$$

As we might expect, the activation energy has units of $\mathrm{J\,mol^{-1}}$. Since $1\,\mathrm{kJ} = 10^3\,\mathrm{J}$ and $10^5 = 10^3 \times 10^2$ we also have

$$E_a = 1.04 \times 10^2\,\mathrm{kJ\,mol^{-1}}$$

$$= 104\,\mathrm{kJ\,mol^{-1}}$$

Worked example 4.18

The activation energy for the reaction

$$2H_2O_2 \rightleftharpoons 2H_2O + O_2$$

is $48.9\,\mathrm{kJ\,mol^{-1}}$ in the presence of a colloidal platinum catalyst. What is the effect on the rate constant of raising the temperature from 20°C to 30°C?

Chemical background

This is another example of a disproportionation reaction. If traces of certain ions are present, these catalyse the reaction which then occurs rapidly. One such ion is Fe^{3+}, which may alternate with Fe^{2+} while the reaction is taking place.

Hydrogen peroxide can be produced industrially by the partial oxidation of 2-ethylanthraquinol or 2-propanol by air. It can be extracted into water.

Solution to worked example

If we assume that the rate constant has a value of k_1 when the absolute temperature is T_1, and a value of k_2 at temperature T_2, we can write the logarithmic form of the Arrhenius equation for the two cases as:

$$\ln k_1 = \ln A - \frac{E_a}{RT_1} \quad \text{and} \quad \ln k_2 = \ln A - \frac{E_a}{RT_2}$$

Subtracting the first of these equations from the second gives

$$\ln k_2 - \ln k_1 = \ln A - \frac{E_a}{RT_2} - \left(\ln A - \frac{E_a}{RT_1}\right)$$

$$= \ln A - \ln A + \frac{E_a}{RT_1} - \frac{E_a}{RT_2}$$

$$= \frac{E_a}{RT_1} - \frac{E_a}{RT_2}$$

$$= \frac{E_a}{R}\left(\frac{1}{T_1} - \frac{1}{T_2}\right)$$

Since

$$\ln X - \ln Y = \ln\left(\frac{X}{Y}\right)$$

as we saw in section 4.4.2, this then becomes

$$\ln\left(\frac{k_2}{k_1}\right) = \frac{E_a}{R}\left(\frac{1}{T_1} - \frac{1}{T_2}\right)$$

and from the left-hand side of the equation we can obtain the ratio k_2/k_1 which indicates the change in the rate constant relative to its initial value k_1.

We need to convert our temperature values in °C to absolute temperature values in K. This gives

$$T_1 = (20 + 273) = 293 \text{ K}$$

and

$$T_2 = (30 + 273) = 303 \text{ K}$$

so that we can now calculate the quantity enclosed in brackets in the above equation which is

$$\left(\frac{1}{T_1} - \frac{1}{T_2}\right) = \frac{1}{293 \text{ K}} - \frac{1}{303 \text{ K}}$$

Evaluating these reciprocals using a calculator gives

$$\left(\frac{1}{T_1} - \frac{1}{T_2}\right) \simeq 3.41 \times 10^{-3}\,\text{K}^{-1} - 3.30 \times 10^{-3}\,\text{K}^{-1}$$

$$= (3.41 - 3.30) \times 10^{-3}\,\text{K}^{-1}$$

$$= 0.11 \times 10^{-3}\,\text{K}^{-1}$$

$$= 1.1 \times 10^{-4}\,\text{K}^{-1}$$

Substituting these values into our equation gives

$$\ln\left(\frac{k_2}{k_1}\right) = \frac{E_a}{R}\left(\frac{1}{T_1} - \frac{1}{T_2}\right)$$

$$= \frac{48.9\,\text{kJ mol}^{-1}}{8.314\,\text{J K}^{-1}\,\text{mol}^{-1}} \times 1.1 \times 10^{-4}\,\text{K}^{-1}$$

$$= \frac{48.9 \times 10^3\,\text{J mol}^{-1}}{8.314\,\text{J K}^{-1}\,\text{mol}^{-1}} \times 1.1 \times 10^{-4}\,\text{K}^{-1}$$

$$\simeq 0.647$$

Notice that the units cancel to give a value for $\ln(k_2/k_1)$ without units, as required. Since

$$\ln\left(\frac{k_2}{k_1}\right) = 0.647$$

we need to take the exponential function of both sides of this equation to give

$$\frac{k_2}{k_1} = e^{0.647}$$

and with the use of a calculator we obtain

$$\frac{k_2}{k_1} = 1.91$$

We see that for this reaction an increase in temperature of 10°C produces almost a doubling of the rate constant.

This reaction has an activation energy which is fairly typical for reactions which proceed at a reasonable rate. A useful rule of thumb, used by photographers, is that a temperature increase of 10°C will produce a doubling of the reaction rate.

Exercises

1. Determine these indefinite integrals.

(a) $\displaystyle\int (6x^2 + 9x + 8)\,dx$

(b) $\displaystyle\int (3x^3 + 4x^2)\,dx$

2. Calculate these definite integrals.

(a) $\displaystyle\int_{-1}^{1} (x^3 + x^2 + x)\, dx$

(b) $\displaystyle\int_{0}^{3} (4x^2 + 2x + 1)\, dx$

3. Determine these integrals.

(a) $\displaystyle\int \left(x + \frac{1}{x} \right) dx$

(b) $\displaystyle\int_{1}^{2} \left(2x^2 + \frac{3}{x} \right) dx$

4. Simplify each of these expressions into a single logarithm:
   (a) $\log 3 + \log 4$
   (b) $\ln 1 + \ln 2 + \ln 3$
5. Express

$$\frac{x}{(x + 2)(x + 3)}$$

in terms of its partial fractions.
6. Differentiate these expressions.
   (a) $2x^2 + 3 \ln x$
   (b) $3x + \ln(x^2 - 2)$
7. Integrate these expressions.

(a) $\displaystyle\int \left(x + \frac{1}{x - 2} \right) dx$

(b) $\displaystyle\int_{0}^{1} \left(2x^2 + \frac{1}{x - 5} \right) dx$

8. Calculate these quantities.
   (a) $2e^{-3}$
   (b) $4e^{2}$
   (c) $3e^{0}$
9. Determine the inverses of these functions.
   (a) $f(x) = 3x^2 + 5$
   (b) $2\sqrt{x} - 8$
10. Evaluate this integral.

$$\int_{0}^{1} \frac{dx}{(x - 4)(x + 5)}$$

Problems

1. Azomethane at an initial pressure of 0.080 mm Hg was allowed to decompose at 350°C. After an hour, its partial pressure was 0.015 mm Hg. What is the average rate of the reaction?

2. Trichloroamine reacts with liquid or concentrated aqueous HCl according to the equation

$$NCl_3 + 4HCl \rightarrow NH_4Cl + 3Cl_2$$

Write down a series of relationships between the rates of change of the concentrations of each of the four chemical species.

3. Determine the overall orders of the following reactions, which have the rate equations shown:

(a) $BrO_3^{-}{}_{(aq)} + 5Br^{-}{}_{(aq)} + 6H^{+}{}_{(aq)} \rightleftharpoons 3Br_{2(aq)} + 3H_2O_{(l)}$

$$\text{rate} = k[BrO_3^{-}][Br^{-}][H^{+}]^2$$

(b) $CH_3COOCH_3 + CH_3CH_2NH_2 \rightleftharpoons CH_3CONHCH_2CH_3 + CH_3OH$

$$\text{rate} = k\frac{[CH_3CH_2NH_2]^{3/2}[CH_3COOCH_3]}{[C_2H_5OH]^{1/2}}$$

4. The concentration of glucose in aqueous hydrochloric acid was monitored and found to give the following results:

$\dfrac{t}{\text{min}}$	$\dfrac{c}{10^2 \text{ mol dm}^{-3}}$
120	5.94
240	5.65
360	5.43
480	5.15

Determine the order of the reaction and the rate constant.

5. Use the Arrhenius equation to calculate the activation energy for a reaction whose rate constant increases by a factor of 3 when the temperature is increased from 20°C to 30°C.

5 Structural chemistry

Although the structural properties of solids, liquids and gases are all important, in this chapter we will be concerned mainly with the structures of solids, and particularly those solids which are crystalline. One of the reasons for this is that far more information is available on the structures of crystalline solids than any other form of matter, this being obtained by the technique of X-ray crystallography.

The concept of symmetry is very important when we are studying structural properties, and we will also look at the use of simple symmetry operators. A more detailed treatment of symmetry involves a reasonable knowledge of group theory, which will not be covered here, since an elementary treatment relevant to chemists is usually included in the appropriate section of physical chemistry textbooks.

5.1 Bragg's Law

One reason for starting a chapter on structural chemistry with a discussion of Bragg's Law is that it is mathematically quite simple, and yet its study introduces many of the techniques which will be of use to us later on. It governs the behaviour of X-rays when they are diffracted from a crystal.

Since crystals consist of units (atoms, ions or molecules) which are arranged in an orderly fashion, they may be represented by a series of lattice points, each of which has the same environment. Various planes may be drawn through these points, as shown in Figure 5.1. Since the spacing of atoms in a crystal is similar to the wavelength of incident X-rays, diffraction rather than reflection occurs with such radiation from planes of atoms. The diffraction angle θ is therefore dependent upon the separation d of particular planes as well as on the wavelength λ of the incident X-rays.

The great advance brought about by the Bragg equation was that it considered radiation of a fixed wavelength. Previously it had only been possible to use 'white' X-rays which were a mixture of various wavelengths and resulted in a far more complicated diffraction pattern.

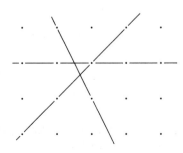

Figure 5.1 Planes in a crystal lattice.

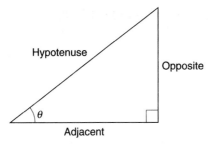

Figure 5.2 Definitions of trigonometric functions.

5.1.1 Trigonometry

The trigonometric functions are simply a set of functions which take as their input an angle. The simplest three are defined in terms of a right-angled triangle containing an angle θ, as shown in Figure 5.2. The lengths of the sides of the triangle are labelled as: opposite the angle of interest; adjacent to the angle of interest; and the hypotenuse (which is opposite the right angle). The sine, cosine and tangent functions are then defined by the equations

$$\sin \theta = \frac{\text{opposite}}{\text{adjacent}}$$

$$\cos \theta = \frac{\text{adjacent}}{\text{hypotenuse}}$$

$$\tan \theta = \frac{\text{opposite}}{\text{adjacent}}$$

which show that each function is defined as the ratio of two lengths, and therefore has no units.

While angles are normally expressed in units of degrees, it is sometimes more convenient to express them in terms of the quantity radians, which has the symbol rad. These are related by the equation

$$2\pi \text{ rad} = 360°$$

so that

$$\pi \text{ rad} = 180°$$

In practice, it is not usually necessary to convert from one to the other since scientific calculators have both 'degrees' and 'radians' mode. Make sure you are working in the correct one! ▦

The graphs of the sine, cosine and tangent functions are shown in Figure 5.3. It is worth remembering that

$$\sin 0° = \cos 90° = 0$$

$$\sin 90° = \cos 0° = 1$$

$$\tan 0° = \frac{\sin 0°}{\cos 0°} = \frac{0}{1} = 0$$

$$\tan 90° = \frac{\sin 90°}{\cos 90°} = \frac{1}{0} \text{ which is undefined}$$

Worked example 5.1

The Bragg equation is expressed as

$$n\lambda = 2d \sin \theta$$

where n is called the order of reflection (having integer values of 1, 2, 3, etc.), and the other terms have been defined above. Calculate the lattice spacing d when copper K_α radiation of wavelength 0.154 nm is incident on a cubic crystal and produces a first-order ($n = 1$) reflection with a scattering angle of 11°.

Chemical background

Other targets used for the generation of X-rays include molybdenum, cobalt, iron and chromium.

X-rays can be generated by firing high energy electrons at a target which is most frequently copper metal. Copper electrons are displaced from the innermost K shell and replaced by those from shells further out such as L and M which release their excess energy as X-rays, as shown in Figure 5.4. When replacement of an electron occurs from the next level up L the radiation is known as K_α, while it is also possible (although less likely) to obtain K_β when the replacement electron is from the next highest level M.

Solution to worked example

To calculate the lattice spacing d, we need to rearrange the equation to make d the subject. We can do this by dividing both sides by $2 \sin \theta$:

$$d = \frac{n\lambda}{2 \sin \theta}$$

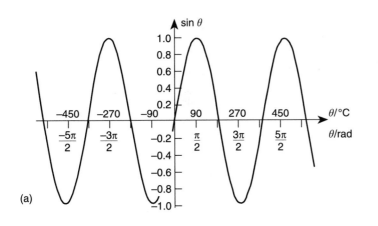

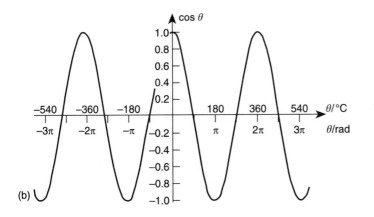

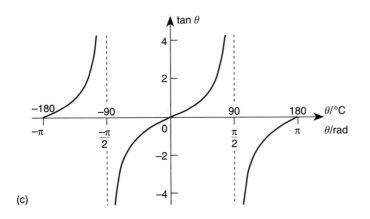

Figure 5.3 Graphs of trigonometric functions: (a) $\sin \theta$; (b) $\cos \theta$; and (c) $\tan \theta$.

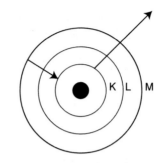

Figure 5.4   Production of K_a X-rays.

We have the quantities required to make a straightforward substitution into this equation giving us

$$d = \frac{1 \times 0.154 \text{ nm}}{2 \times \sin 11°}$$

$$= \frac{0.154 \text{ nm}}{2 \times 0.1908}$$

$$\simeq 0.404 \text{ nm}$$

Notice that in working through this problem we did not need to convert the units of the wavelength from nm to m, and the units of the spacing of the planes follow directly from substituting into the equation.

5.1.2   Inverses of trigonometric functions

The idea of inverse functions was introduced in section 4.6.2. Each of the trigonometric functions we have met has its own inverse. These are

$$\arcsin x \text{ written as } \sin^{-1} x$$

$$\arccos x \text{ written as } \cos^{-1} x$$

$$\arctan x \text{ written as } \tan^{-1} x$$

The use of -1 as a superscript in these expressions does *not* have any relationship to its use elsewhere to represent a reciprocal, and is purely the nomenclature used to denote the inverse function.

To see how the inverse trigonometric functions are used in practice, consider the function defined as

$$f(x) = \sin(3x + 4)$$

When calculating the inverse of this function, we clearly need to generate x from $\sin x$ and this can only be done by using the inverse trigonometric function. We must therefore have the term $\sin^{-1} x$ in the expression for the inverse function arc $f(x)$. Having done that, we simply work the operations in reverse order. In this case, that involves subtracting 4 and then dividing by 3 so that we obtain

$$\text{arc } f(x) = \frac{(\sin^{-1} x) - 4}{3}$$

Worked example 5.2

A certain set of lattice planes in potassium nitrate crystals has a spacing of 543 pm. Calculate the first-order ($n = 1$) diffraction angle when copper K_α radiation of wavelength 154 pm is incident on these planes.

Potassium nitrate can be prepared by fractional crystallization from a solution of sodium nitrate and potassium chloride. It is used in gunpowder.

Chemical background

The crystal structure of potassium nitrate consists of an orthorhombic lattice in which the angles are each 90°. The dimensions of the sides of the unit cell are 5.431 Å, 9.164 Å and 5.414 Å respectively.

Solution to worked example

It is probably easiest to determine the value of $\sin \theta$ and then to generate θ by using the inverse function. The first stage is to rearrange the Bragg equation to make $\sin \theta$ the subject. This can be done by dividing both sides by $2d$:

$$\sin \theta = \frac{n\lambda}{2d}$$

which on substitution of the values given leads to

$$\sin \theta = \frac{1 \times 154 \text{ pm}}{2 \times 543 \text{ pm}}$$

$$\simeq 0.1418$$

We now need to apply the inverse sine function to both sides of this equation.

$$\sin^{-1}(\sin \theta) = \sin^{-1}(0.1418) \quad \text{▦}$$

Notice that $\sin^{-1}(\sin \theta) = \theta$. Using a calculator for $\sin^{-1}(0.1418)$ we find

$$\theta \simeq 8.15°$$

The lattice points referred to in section 5.1 can be used to define the unit cell of a crystal. This is simply a three-dimensional unit from which it is possible to

5.2
The unit cell

Figure 5.5 Example of a crystal lattice in two dimensions.

The monoclinic form of sulphur crystallizes in the space group known as $P2_1/c$.

generate the whole lattice (Figure 5.5). Unit cells may be primitive, having lattice points only at their corners, or non-primitive in which case additional lattice points are present.

5.2.1 Unit vectors

Vectors are simply numbers which are associated with a particular direction. Because of this they are very useful for dealing with problems in three dimensions. Any vector can be defined in terms of its components in three directions, normally chosen to be mutually at right angles and denoted as x, y and z. **Unit vectors**, having a size of one unit, can be defined in each of the directions and are denoted as **i**, **j** and **k** respectively. Note the use of bold typescript to denote vector quantities; when written by hand they can simply be underlined. These unit vectors can be combined by simple addition, so that a vector **a** might be defined as

$$\mathbf{a} = 2\mathbf{i} + 3\mathbf{j} + \mathbf{k}$$

This vector is shown relative to the three axes x, y and z in Figure 5.6.

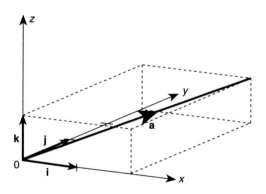

Figure 5.6 The vector $2\mathbf{i} + 3\mathbf{j} + \mathbf{k}$.

This defines a vector which can be thought of as starting at the origin, and whose finish is found by moving 2 units in the *x*-direction, 3 units in the *y*-direction and 1 unit in the *z*-direction. The actual vector is then represented diagrammatically by a straight line starting at the origin and finishing at this second point, as in Figure 5.6.

5.2.2 Addition and subtraction of vectors

Vectors can be added and subtracted using the standard rules of arithmetic applied to each direction separately. For example, if vectors **b** and **c** are defined as

$$\mathbf{b} = 2\mathbf{i} + 3\mathbf{j}$$

$$\mathbf{c} = 2\mathbf{j} + \mathbf{k}$$

then

$$\mathbf{b} + \mathbf{c} = 2\mathbf{i} + 3\mathbf{j} + 2\mathbf{j} + \mathbf{k}$$

$$= 2\mathbf{i} + (3\mathbf{j} + 2\mathbf{j}) + \mathbf{k}$$

$$= 2\mathbf{i} + 5\mathbf{j} + \mathbf{k}$$

Worked example 5.3

Figure 5.7 shows lattice points situated at the corners of a primitive orthogonal lattice having dimensions *a*, *b* and *c*; the origin O and axes *x*, *y* and *z* are shown. In terms of the unit vectors **i**, **j** and **k**, determine the position vector of each point A–G.

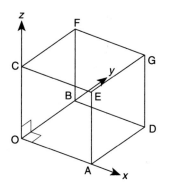

Figure 5.7 Primitive orthogonal lattice.

Chemical background

Such a crystal lattice, containing points only at the corners of each unit cell, is known as a primitive lattice. Because it has the form of a cuboid if *a*, *b* and *c* are different, it is known as orthorhombic. An example of a structure having a primitive orthorhombic unit cell is that of the mineral Forsterite, $Mg_2(SiO_4)$.

Solution to worked example

We need to consider each lattice point in turn, and to define the path which takes us from the origin to that point.

(a) To move from O to A, we need to go *a* units in the *x*-direction. Since the unit vector in this direction is **i**, the vector we require is *a***i**.
(b) Point B lies on the *y*-axis at a distance of *b* units from O. The unit vector along this axis is **j**, so the required vector is *b***j**.
(c) Point C lies on the *z*-axis, along which the unit vector is **k**. We need to move *c* units, so this point is defined by vector *c***k**.
(d) To get from O to D we need to move *a* units in the *x*-direction and *b* units in the *y*-direction. The vector is consequently *a***i** + *b***j**.
(e) Point E is reached by moving *a* units in the *x*-direction and *c* units in the *z*-direction. No movement in the *y*-direction is required. The defining vector is therefore *a***i** + *c***k**.
(f) Point F requires movement of *b* units in the *y*-direction and *c* units in the *z*-direction, so the required vector is *b***j** + *c***k**.
(g) Point G is the only one which requires a movement in each of the three directions; *a* units in *x*, *b* units in *y* and *c* units in *z*. The vector defining this point is consequently *a***i** + *b***j** + *c***k**.

5.2.3 Multiplication of vectors

There are two ways in which a pair of vectors can be multiplied together. One results in a pure number, also known as a scalar quantity, while the other gives another vector. The first method is known as calculating the **scalar product**, or dot product since this is the symbol used to denote the process. For example, the scalar product of two vectors **a** and **b** would be denoted as **a** · **b** and is defined as

$$\mathbf{a} \cdot \mathbf{b} = |\mathbf{a}||\mathbf{b}| \cos \theta$$

where |**a**| and |**b**| are known as the **moduli** of **a** and **b** respectively and θ is the angle between them.

   The modulus or magnitude of a vector is its length (a pure number or scalar quantity) and can be calculated by taking the square root of the sums of the

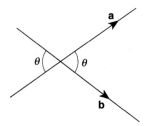

Figure 5.8   Definition of the angle between two vectors.

squares of the contributions of each unit vector to the total. In two dimensions, this is Pythagoras' Theorem: $c = \sqrt{(a^2 + b^2)}$. In three dimensions, for example, if

$$\mathbf{v} = 2\mathbf{i} + \mathbf{j} - 3\mathbf{k}$$

then

$$\begin{aligned}|\mathbf{v}| &= \sqrt{(2^2 + 1^2 + (-3)^2)} \\ &= \sqrt{(4 + 1 + 9)} \\ &= \sqrt{14}\end{aligned}$$

There are two possible angles which can be taken as θ, since $\cos\theta = \cos(180 - \theta)$ and your calculator will offer the acute angle θ. The correct one is shown in Figure 5.8 which shows that both vectors must be pointing away from or both must be pointing towards the point of intersection. The cosine of this angle, known as $\cos\theta$ is easily calculated using a scientific calculator, but it is important to ensure that it has been set to the appropriate mode to accept angles in degrees as input, and that you choose the correct angle θ, or $180 - \theta$.

Notice also that $\mathbf{a} \cdot \mathbf{b} = \mathbf{b} \cdot \mathbf{a}$; it does not matter which order you take dot products.

If we consider the scalar products of the various combinations of unit vectors we obtain a very useful set of relationships. For example,

$$\mathbf{i} \cdot \mathbf{i} = |\mathbf{i}||\mathbf{i}| \cos\theta$$

Now $|\mathbf{i}| = 1$ and the angle between $\mathbf{i}$ and $\mathbf{i}$ is zero so $\theta = 0°$. Since $\cos 0° = 1$ we then have

$$\mathbf{i} \cdot \mathbf{i} = 1$$

The same set of arguments leads to the relationships

$$\mathbf{j} \cdot \mathbf{j} = 1 \quad \text{and} \quad \mathbf{k} \cdot \mathbf{k} = 1$$

On the other hand, if we consider a scalar product such as $\mathbf{i} \cdot \mathbf{j}$ we then have

$$\mathbf{i} \cdot \mathbf{j} = |\mathbf{i}||\mathbf{j}| \cos\theta$$

where again $|\mathbf{i}| = 1$ and $|\mathbf{j}| = 1$ but this time the angle between $\mathbf{i}$ and $\mathbf{j}$ is a right angle so $\theta = 90°$. Since $\cos 90° = 0$, we now have $\mathbf{i} \cdot \mathbf{j} = 0$. Notice then that

$$\mathbf{i} \cdot \mathbf{j} = \mathbf{j} \cdot \mathbf{i} = 0$$

Similarly

$$\mathbf{j} \cdot \mathbf{k} = \mathbf{k} \cdot \mathbf{j} = 0$$

and

$$\mathbf{i} \cdot \mathbf{k} = \mathbf{k} \cdot \mathbf{i} = 0$$

Now consider what happens if we need to calculate the scalar product of two vectors which are defined in terms of the unit vectors $\mathbf{i}$, $\mathbf{j}$ and $\mathbf{k}$, such as

$$\mathbf{a} = 3\mathbf{i} + 2\mathbf{j} + \mathbf{k} \quad \text{and} \quad \mathbf{b} = 2\mathbf{i} - \mathbf{j} + \mathbf{k}$$

We can expand this product as we would with any pair of brackets:

$$\mathbf{a} \cdot \mathbf{b} = (3\mathbf{i} + 2\mathbf{j} + \mathbf{k}) \cdot (2\mathbf{i} - \mathbf{j} + \mathbf{k})$$

$$= 3\mathbf{i} \cdot (2\mathbf{i} - \mathbf{j} + \mathbf{k})$$

$$+ 2\mathbf{j} \cdot (2\mathbf{i} - \mathbf{j} + \mathbf{k})$$

$$+ \mathbf{k} \cdot (2\mathbf{i} - \mathbf{j} + \mathbf{k})$$

$$= 6\mathbf{i} \cdot \mathbf{i} - 3\mathbf{i} \cdot \mathbf{j} + 3\mathbf{i} \cdot \mathbf{k}$$

$$+ 4\mathbf{j} \cdot \mathbf{i} - 2\mathbf{j} \cdot \mathbf{j} + 2\mathbf{j} \cdot \mathbf{k}$$

$$+ 2\mathbf{k} \cdot \mathbf{i} - \mathbf{k} \cdot \mathbf{j} + \mathbf{k} \cdot \mathbf{k}$$

Since

$$\mathbf{i} \cdot \mathbf{j} = \mathbf{j} \cdot \mathbf{i} = \mathbf{j} \cdot \mathbf{k} = \mathbf{k} \cdot \mathbf{j} = \mathbf{i} \cdot \mathbf{k} = \mathbf{k} \cdot \mathbf{i} = 0$$

and

$$\mathbf{i} \cdot \mathbf{i} = \mathbf{j} \cdot \mathbf{j} = \mathbf{k} \cdot \mathbf{k} = 1$$

this simply leaves us with

$$\mathbf{a} \cdot \mathbf{b} = 6 - 2 + 1$$

$$= 5$$

As well as calculating the scalar product of a pair of vectors, there are occasions when we need to multiply vectors and retain some of the information about direction. To do this, we can calculate the **vector product** of a pair of vectors. This is denoted as $\mathbf{a} \times \mathbf{b}$, and is defined as

$$\mathbf{a} \times \mathbf{b} = |\mathbf{a}||\mathbf{b}| \sin \theta \hat{\mathbf{n}}$$

where the symbols have the same meaning as in the case of the scalar product. Notice this time, however, that the vector product is defined in terms of a vector $\hat{\mathbf{n}}$.

This simply gives the direction of the resulting vector, and is itself of magnitude 1. $\hat{\mathbf{n}}$ is perpendicular to both the vectors $\mathbf{a}$ and $\mathbf{b}$, and its direction (up or down) can be found by using the corkscrew rule. If we imagine a right handed corkscrew which rotates from vector $\mathbf{a}$ to vector $\mathbf{b}$, then the direction of travel of the corkscrew gives the direction of the vector $\hat{\mathbf{n}}$.

Note that $\mathbf{a} \times \mathbf{b} \neq \mathbf{b} \times \mathbf{a}$ because the corkscrew rule results in the vector product pointing in opposite directions. However

$$\mathbf{a} \times \mathbf{b} = -\mathbf{b} \times \mathbf{a}$$

We can apply this rule to calculate the vector products between pairs of the unit vectors $\mathbf{i}$, $\mathbf{j}$ and $\mathbf{k}$. Since the angle between pairs of unit vectors is $0°$, $\sin \theta = 0$, and so the vector products between a pair of identical vectors must be zero. This gives

$$\mathbf{i} \times \mathbf{i} = \mathbf{j} \times \mathbf{j} = \mathbf{k} \times \mathbf{k} = 0$$

where the zero is technically a vector quantity. We also know that $\sin 90°$ is 1, so for a pair of perpendicular vectors we do not need to worry about this term and we again obtain a unit vector. If $\mathbf{i}$, $\mathbf{j}$ and $\mathbf{k}$ form a right-handed set, so that the corkscrew rule can be applied, we obtain

$$\mathbf{i} \times \mathbf{j} = \mathbf{k}$$

$$\mathbf{k} \times \mathbf{i} = \mathbf{j}$$

$$\mathbf{j} \times \mathbf{k} = \mathbf{i}$$

Since $\mathbf{a} \times \mathbf{b} = -\mathbf{b} \times \mathbf{i}$, we also have

$$\mathbf{j} \times \mathbf{i} = -\mathbf{k}$$

$$\mathbf{i} \times \mathbf{k} = -\mathbf{j}$$

$$\mathbf{k} \times \mathbf{j} = -\mathbf{i}$$

This information allows us to calculate the vector product of any pair of vectors which are expressed in terms of the unit vectors $\mathbf{i}$, $\mathbf{j}$ and $\mathbf{k}$. Using the example above, where

$$\mathbf{a} = 3\mathbf{i} + 2\mathbf{j} + \mathbf{k}$$

$$\mathbf{k} = 2\mathbf{i} - \mathbf{j} + \mathbf{k}$$

the expansion of the brackets gives

$$\mathbf{a} \times \mathbf{b} = (3\mathbf{i} + 2\mathbf{j} + \mathbf{k}) \times (2\mathbf{i} - \mathbf{j} + \mathbf{k})$$

$$= 3\mathbf{i} \times (2\mathbf{i} - \mathbf{j} + \mathbf{k})$$

$$+ 2\mathbf{j} \times (2\mathbf{i} - \mathbf{j} + \mathbf{k})$$

$$+ \mathbf{k} \times (2\mathbf{i} - \mathbf{j} + \mathbf{k})$$

$$= 3\mathbf{i} \times 2\mathbf{i} - 3\mathbf{i} \times \mathbf{j} + 3\mathbf{i} \times \mathbf{k}$$

$$+ 2\mathbf{j} \times 2\mathbf{i} - 2\mathbf{j} \times \mathbf{j} + 2\mathbf{j} \times \mathbf{k}$$

$$+ 2\mathbf{k} \times \mathbf{i} - \mathbf{k} \times \mathbf{j} + \mathbf{k} \times \mathbf{k}$$

Now

$$\mathbf{i} \times \mathbf{i} = \mathbf{j} \times \mathbf{j} = \mathbf{k} \times \mathbf{k} = 0$$

This leaves

$$\mathbf{a} \times \mathbf{b} = -3\mathbf{i} \times \mathbf{j} + 3\mathbf{i} \times \mathbf{k}$$
$$+ 4\mathbf{j} \times \mathbf{i} + 2\mathbf{j} \times \mathbf{k}$$
$$+ 2\mathbf{k} \times \mathbf{i} - \mathbf{k} \times \mathbf{j}$$

We can now substitute for each of these individual vector products to obtain

$$\mathbf{a} \times \mathbf{b} = -3\mathbf{k} + 3(-\mathbf{j}) + 4(-\mathbf{k}) + 2\mathbf{i} + 2\mathbf{j} - (-\mathbf{i})$$
$$= -3\mathbf{k} - 3\mathbf{j} - 4\mathbf{k} + 2\mathbf{i} + 2\mathbf{j} + \mathbf{i}$$
$$= 3\mathbf{i} - \mathbf{j} - 7\mathbf{k}$$

Worked example 5.4

The methane molecule may be thought of as containing a carbon atom at the centre of a cuboid together with four hydrogen atoms at alternate corners. Use this information to calculate the bond angle in methane.

Chemical background

The tetrahedral structure of methane was also predicted by van't Hoff in 1874, on the basis of the observation that only one isomer of compounds such as CH_3Cl and CH_3Br has ever been observed.

Using the Valence Shell Electron Pair Repulsion (VSEPR) model, we would expect the four regions of electron density around the central carbon atom to be as far away from each other as possible. This leads to the arrangement described in the question, which is a regular tetrahedron. Such a structure has been confirmed experimentally by means of electron diffraction.

Solution to worked example

It is necessary to formulate this problem in terms of unit vectors. If we set up the cuboid as shown in Figure 5.9 we can define three axes x, y and z as shown. The carbon atom C is placed at the centre and the four hydrogen atoms at alternate corners. We choose to label two of these A and B and will be calculating the angle A-C-B. Atom A is chosen to be at the origin of our coordinate system, and we will assign the cuboid an arbitrary length d. We are going to determine the angle by calculating the scalar product between vectors $\mathbf{AC}$ and $\mathbf{BC}$.

We can define each of these vectors in terms of the unit vectors $\mathbf{i}$, $\mathbf{j}$ and $\mathbf{k}$ by referring to Figure 5.9. One way of moving from A to C is to move along half the side of the cube in the x-direction, then by the same amount in the y-direction, and finally by the same amount again in the z-direction. This allows us to define the vector from A to C as

$$\mathbf{AC} = \tfrac{1}{2}d\mathbf{i} + \tfrac{1}{2}d\mathbf{j} + \tfrac{1}{2}d\mathbf{k}$$

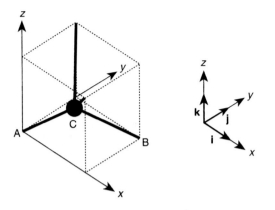

Figure 5.9   Carbon atom in a tetrahedral environment.

To move from B to C, we need to move $\frac{1}{2}d$ in the negative x-direction, $\frac{1}{2}d$ in the negative y-direction and $\frac{1}{2}d$ in the positive z-direction. We therefore have

$$\mathbf{BC} = -\tfrac{1}{2}d\mathbf{i} - \tfrac{1}{2}d\mathbf{j} + \tfrac{1}{2}d\mathbf{k}$$

Both of these vectors will have the same modulus, which is

$$|\mathbf{AC}| = |\mathbf{BC}| = \sqrt{\left(\left(\tfrac{1}{2}d\right)^2 + \left(\tfrac{1}{2}d\right)^2 + \left(\tfrac{1}{2}d\right)^2\right)}$$
$$= d\sqrt{\tfrac{3}{4}}$$

and so the scalar product is

$$\mathbf{AC} \cdot \mathbf{BC} = d\sqrt{\tfrac{3}{4}} \times d\sqrt{\tfrac{3}{4}} \times \cos\theta$$
$$= d^2 \times \tfrac{3}{4} \times \cos\theta$$
$$= \tfrac{3}{4}d^2 \cos\theta$$

where θ is the required bond angle.

Calculation of the scalar product by combining the unit vectors and using

$$\mathbf{i} \cdot \mathbf{i} = \mathbf{j} \cdot \mathbf{j} = \mathbf{k} \cdot \mathbf{k} = 1$$

and

$$\mathbf{i} \cdot \mathbf{j} = \mathbf{j} \cdot \mathbf{k} = \mathbf{k} \cdot \mathbf{i} = 0$$

gives

$$\mathbf{AC} \cdot \mathbf{BC} = \left(\tfrac{1}{2}d\right)\left(-\tfrac{1}{2}d\right) + \left(\tfrac{1}{2}d\right)\left(-\tfrac{1}{2}d\right) + \left(\tfrac{1}{2}d\right)\left(\tfrac{1}{2}d\right)$$
$$= -\tfrac{1}{4}d^2 - \tfrac{1}{4}d^2 + \tfrac{1}{4}d^2$$
$$= -\tfrac{1}{4}d^2$$

These two expressions for the scalar product can now be equated to give

$$\tfrac{3}{4}d^2 \cos\theta = -\tfrac{1}{4}d^2$$

Dividing both sides by d^2 and multiplying both sides by 4 gives

$$3 \cos \theta = -1$$

$$\cos \theta = -\tfrac{1}{3}$$

and so θ can be found by taking the inverse function:

$$\theta = \cos^{-1} -\tfrac{1}{3}$$

Using a calculator set to 'degrees' mode ▦ gives

$$\theta \simeq 109.5°$$

which is the tetrahedral value expected.

Worked example 5.5

A unit cell can be defined in terms of the three vectors **a**, **b** and **c** along its edges. The angle between **b** and **c** is known as α, that between **a** and **c** as β and that between **a** and **b** as γ. A monoclinic crystal is defined as one having α and γ equal to 90° with β having a value of greater than 90°. In this case, the vectors **a**, **b** and **c** are related to the unit vectors **i**, **j** and **k** by the equations

$$\mathbf{a} = a \sin \beta \mathbf{i} + a \cos \beta \mathbf{k}$$

$$\mathbf{b} = b\mathbf{j}$$

$$\mathbf{c} = c\mathbf{k}$$

where a, b and c are simply the magnitudes of the vectors **a**, **b** and **c** respectively. Use the fact that the volume V of the unit cell is given by the triple scalar product

$$V = \mathbf{a} \cdot (\mathbf{b} \times \mathbf{c})$$

to obtain an expression for the volume of the unit cell in terms of a, b, c and β.

Chemical background

Unit cells can be characterized into various types on the basis of the relationships between the dimensions a, b, c, α, β and γ. The most general case, in which these quantities are all independent, is known as a triclinic unit cell. In the hexagonal cell a and b are equal, α and β are both 90° and γ is 120°.

Solution to worked example

This question requires us to calculate a **triple scalar product**; it is called this because three vectors are involved and the result will be a scalar quantity.

We evaluate the vector product enclosed in brackets first, and then use this result to calculate the scalar product.

The vector product is given by

$$\mathbf{b} \times \mathbf{c} = b\mathbf{j} \times c\mathbf{k}$$

$$= bc\mathbf{j} \times \mathbf{k}$$

since the scalar quantities b and c can be multiplied directly. Since $\mathbf{j} \times \mathbf{k} = \mathbf{i}$, we now have

$$\mathbf{b} \times \mathbf{c} = bc\mathbf{i}$$

We are now able to calculate the required scalar product, which is given by

$$\mathbf{a} \cdot (\mathbf{b} \times \mathbf{c}) = (a \sin \beta \mathbf{i} + a \cos \beta \mathbf{k}) \cdot (bc\mathbf{i})$$

$$= (a \sin \beta \mathbf{i}) \cdot (bc\mathbf{i}) + (a \cos \beta \mathbf{k}) \cdot (bc\mathbf{i})$$

$$= abc \sin \beta \mathbf{i} \cdot \mathbf{i} + abc \cos \beta \mathbf{k} \cdot \mathbf{i}$$

simply multiplying out the terms in the brackets. Since $\mathbf{k} \cdot \mathbf{i} = 0$ and $\mathbf{i} \cdot \mathbf{i} = 1$ this leaves

$$\mathbf{a} \cdot (\mathbf{b} \times \mathbf{c}) = abc \sin \beta$$

which is equal to the volume of the monoclinic unit cell.

5.3 X-ray diffraction

Bragg's Law gives us information about the behaviour of diffracted X-rays from one particular set of planes in a crystal. However, as we have seen above, there are many planes existing in a crystal lattice. Interpretation of the complex patterns, arising from the diffraction of X-rays from many different crystal planes simultaneously, enables us to obtain detailed information on crystal structure. It is generally possible to locate the positions of atoms in a unit cell, and these positions are generally expressed in terms of fractions of each of the dimensions a, b and c, known as fractional crystallographic coordinates and often written as x/a, y/b and z/c.

5.3.1 Complex numbers

A complex number z is one of the form

$$z = a + ib$$

where a and b are 'real' numbers, which may be positive or negative and have values such as -3.4, 2.6, 7.0 and so on. The number i is defined by the equation

$$i^2 = -1$$

which seems strange at first when you realize that taking the square root of either side leads to

$$i = \sqrt{-1}$$

This i should not be confused with the unit vector **i** which we met earlier in this chapter. You may also see the square root of -1 represented by the symbol j, but this could also be confused with a unit vector.

In a complex number z, the number a is known as the **real** part of z, and b is known as the **imaginary** part of z. So, if we have the complex number

$$z = 2 - 3i$$

its real part is 2 and its imaginary part -3.

In chemistry, such numbers have uses in crystallography and also in quantum mechanics, as we will see in Chapter 6.

We can perform the usual arithmetic operations on complex numbers. In the case of addition and subtraction, we need to treat the real and imaginary parts separately, so that for example

$$(3 - 2i) + (5 + i) = (3 + 5) + (-2i + i)$$

$$= 8 - i$$

In the case of multiplication, we simply consider every possible term as we have seen previously when multiplying a pair of brackets. For example

$$(2 - 3i)(4 + i) = 2 \times (4 + i) - 3i \times (4 + i)$$

$$= 8 + 2i - 12i - 3i^2$$

Since $i^2 = -1$, this becomes

$$8 - 10i - 3(-1) = 8 - 10i + 3$$

$$= 11 - 10i$$

There is a useful relationship regarding the exponential of an imaginary number, one which contains only an imaginary part. This is

$$e^{ikx} = \cos kx + i \sin kx$$

$$e^{-ikx} = \cos kx - i \sin kx$$

Worked example 5.6

The structure factor $F(h \quad k \quad l)$ for a particular reflection of X-rays from a crystal is made of contributions from each of the atoms (x_j, y_j, z_j) in a unit cell, and is governed by the values of the integers h, k and l. The expression for its calculation is

$$F(h \quad k \quad l) = \sum f_j \exp(2\pi i(hx_j + ky_j + lz_j))$$

where f_j is the atomic scattering factor for atom j. In a face-centred cubic lattice, the atomic positions (x_j, y_j, z_j) are $(0, 0, 0)$, $(\frac{1}{2}, \frac{1}{2}, 0)$, $(\frac{1}{2}, 0, \frac{1}{2})$ and $(0, \frac{1}{2}, \frac{1}{2})$. Show that if h, k and l are all even or all odd $F(h \quad k \quad l) = 4f_j$ but otherwise $F(h \quad k \quad l) = 0$.

Chemical background

The structure factor gives the amplitude of all the diffracted waves when the contributions from each atom in the unit cell are added together. The square of this is then proportional to the intensity which can be measured. The atomic scattering factor is a measure of how strongly an atom diffracts and, since the electrons are responsible for diffraction, it is proportional to the number of electrons in an atom of a given type.

The integers h, k and l define particular planes in the crystal and, as this problem shows, certain combinations of these values lead to zero structure factors and hence intensities for certain symmetries. In fact, these systematic absences, as they are known, are characteristic of particular space groups.

International Tables for Crystallography list the reflection conditions for the space group $P2_1/c$ as

$h0l$	$l = 2n$
$0k0$	$k = 2n$
$00l$	$l = 2n$

Thus, for odd values of the indices specified, no reflections will be seen.

Solution to worked example

The first stage is to expand the expression given for $F(h \quad k \quad l)$ in terms of the cos and sin functions. This gives

$$F(h \quad k \quad l) = \sum f_j \exp(2\pi i(hx_j + ky_j + lz_j))$$
$$= \sum f_j[\cos 2\pi(hx_j + ky_j + lz_j) + i \sin 2\pi(hx_j + ky_j + lz_j)]$$

Remembering that the symbol $\sum$ denotes summation, we now need to write this expression out in full by substituting the values of (x_j, y_j, z_j) for each of the four positions given. We do this and obtain

$$F(h \quad k \quad l) = f_j[\cos 0 + i \sin 0]$$
$$+ f_j[\cos 2\pi(\tfrac{1}{2}h + \tfrac{1}{2}k) + i \sin 2\pi(\tfrac{1}{2}h + \tfrac{1}{2}k)]$$
$$+ f_j[\cos 2\pi(\tfrac{1}{2}h + \tfrac{1}{2}l) + i \sin 2\pi(\tfrac{1}{2}h + \tfrac{1}{2}l)]$$
$$+ f_j[\cos 2\pi(\tfrac{1}{2}k + \tfrac{1}{2}l) + i \sin 2\pi(\tfrac{1}{2}k + \tfrac{1}{2}l)]$$

This can be simplified since, for example,

$$\cos 2\pi(\tfrac{1}{2}h + \tfrac{1}{2}k) = \cos \pi(h + k)$$

and we know that $\cos 0 = 1$ and $\sin 0 = 0$, so we obtain

$$F(h \quad k \quad l) = f_j + f_j[\cos \pi(h + k) + i \sin \pi(h + k)]$$
$$+ f_j[\cos \pi(h + l) + i \sin \pi(h + l)]$$
$$+ f_j[\cos \pi(k + l) + i \sin \pi(k + l)]$$

If we look at Figure 5.3(a), we see that the value of $\sin \theta$ is zero when $\theta = -2\pi$, $-\pi$, 0, π, 2π. These are multiples of π, so all of the sine terms in the above equation will be zero since h, k and l are all integers. This leaves us with

$$F(h \quad k \quad l) = \sum [1 + \cos \pi(h + k) + \cos \pi(h + l) + \cos \pi(k + l)]$$

If we now look at Figure 5.3(b), we see that $\cos \theta$ is 1 when θ has the values -2π, 0 and 2π, in other words at multiples of 2π. When θ has values of $-\pi$ and π it is equal to -1, in other words at odd multiples of π.

We have four cases to consider. First, if h, k and l are all even then the quantities $\pi(h + k)$, $\pi(h + l)$ and $\pi(k + l)$ will all be even multiples of π and so we have

$$F(h \quad k \quad l) = f_j[1 + 1 + 1 + 1] = 4f_j$$

Equally, if h, k and l are all odd, then the quantities $\pi(h + k)$, $\pi(h + l)$ and $\pi(k + l)$ will all be even multiples of π and so, as above,

$$F(h \quad k \quad l) = 4f_j$$

The situation is more complicated if one or two of h, k and l are odd. For example, if h is odd and k and l are both even then we have

$$h + k \text{ is odd}$$

$$h + l \text{ is odd}$$

$$k + l \text{ is even}$$

Since odd multiples of π give $\cos \theta = -1$ and even multiples of π give $\cos \theta = 1$, this will give

$$F(h \quad k \quad l) = f_j[1 - 1 - 1 + 1] = 0$$

The same reasoning applies if k or l is odd and the other two are even.

Now suppose two of h, k and l are odd. If both h and k are odd, we have

$$h + k \text{ is even}$$

$$h + l \text{ is odd}$$

$$k + l \text{ is odd}$$

which leads to

$$F(h \quad k \quad l) = f_j[1 + 1 - 1 - 1] = 0$$

The same applies for other pairs of odd values of h, k and l.

5.4
Symmetry operators

There are several types of symmetry elements, which include mirror planes, rotation axes and centres of symmetry.

We have already met the idea that crystals consist of regular repeating units. Another way of expressing this is to say that they possess symmetry. As well as being important in crystals, symmetry may also be evident at the molecular level. For example, when we considered methane in Worked Example 5.4 it was obvious that all the hydrogen atoms were equivalent and that we could choose any pair when calculating the bond angle. In fact, if you were to rotate the methane molecule by 120° about any bond then you would obtain a molecule which was indistinguishable in orientation from the one you started with.

The subject of symmetry is quite complex, and a thorough understanding requires some knowledge of the branch of mathematics known as group theory. Here we will only look at how individual symmetry operations can be represented.

5.4.1 Matrices

A matrix is a set of numbers arranged in rows and columns, and enclosed by one pair of brackets per matrix. An example would be

$$\begin{pmatrix} 2 & 3 \\ 4 & 5 \end{pmatrix}$$

This is known as a 2×2 matrix since it has 2 rows and 2 columns. A 2×3 matrix has 2 rows and 3 columns and an example would be

$$\begin{pmatrix} 1 & 2 & 3 \\ 7 & 8 & 9 \end{pmatrix}$$

We can only calculate the sum or difference of two matrices if they have the same dimensions. The operation is simply performed using the corresponding elements of the two matrices. For example, if

$$\mathbf{A} = \begin{pmatrix} 2 & 3 \\ 4 & 5 \end{pmatrix} \quad \text{and} \quad \mathbf{B} = \begin{pmatrix} 1 & 2 \\ 3 & 4 \end{pmatrix}$$

we have

$$\begin{aligned} \mathbf{A} + \mathbf{B} &= \begin{pmatrix} 2 & 3 \\ 4 & 5 \end{pmatrix} + \begin{pmatrix} 1 & 2 \\ 3 & 4 \end{pmatrix} \\ &= \begin{pmatrix} 2+1 & 3+2 \\ 4+3 & 5+4 \end{pmatrix} \\ &= \begin{pmatrix} 3 & 5 \\ 7 & 9 \end{pmatrix} \end{aligned}$$

and

$$\begin{aligned} \mathbf{A} - \mathbf{B} &= \begin{pmatrix} 2 & 3 \\ 4 & 5 \end{pmatrix} - \begin{pmatrix} 1 & 2 \\ 3 & 4 \end{pmatrix} \\ &= \begin{pmatrix} 2-1 & 3-2 \\ 4-3 & 5-4 \end{pmatrix} \\ &= \begin{pmatrix} 1 & 1 \\ 1 & 1 \end{pmatrix} \end{aligned}$$

It is also possible to multiply a matrix by a number, which effectively acts as a scaling factor. For the example above, we could scale matrix **A** by a factor of 2 so that

$$2\mathbf{A} = 2\begin{pmatrix} 2 & 3 \\ 4 & 5 \end{pmatrix} = \begin{pmatrix} 4 & 6 \\ 8 & 10 \end{pmatrix}$$

Two matrices can only be multiplied together if the number of columns in the first matrix is the same as the number of rows in the second. It is possible for a 3×2 matrix to be multiplied by a 2×1 matrix; the result would be a 3×1 matrix. However, you cannot multiply a 2×1 matrix by a 3×2 matrix because the number of columns in the first (1) does not match the number of rows in the second matrix (3). The order of multiplying matrices is therefore important.

To perform the multiplication, we need to multiply each element in the rows of the first matrix by the corresponding elements in the columns of the second matrix. These pairs of numbers are then added to give the element of the matrix which appears in the position given by the intersection of this row and column. For example, if we wish to calculate

$$\begin{pmatrix} 2 & 4 \\ 6 & 8 \\ 7 & 1 \end{pmatrix}\begin{pmatrix} 1 & 2 \\ 5 & 6 \end{pmatrix}$$

then since we are multiplying a 3×2 matrix by a 2×2 matrix we expect to obtain a 3×2 matrix. Matching elements from the *first* row of the first matrix and the *first* column of the second matrix gives us

$$(2 \times 1) + (4 \times 5) = 22$$

This result appears in the position defined by the intersection, i.e. the *first* row and the *first* column in the results matrix. Now working along the *second* row of the first matrix and the *first* column of the second matrix, we have

$$(6 \times 1) + (8 \times 5) = 46$$

and this appears in the position defined by the first column and the second row of the results matrix. We build up the whole matrix in this way to give us the final result:

$$\begin{pmatrix} 22 & 28 \\ 46 & 60 \\ 12 & 20 \end{pmatrix}$$

Worked example 5.7

The operations defined by the symmetry elements in methane can be expressed in matrix notation. The four hydrogen atoms can be represented by the 1×4

matrix

$$(H_A \quad H_B \quad H_C \quad H_D)$$

Multiply this matrix by each of the following 4×4 symmetry matrices to determine the 1×4 matrix which defines the resulting position of each hydrogen atom:

It is possible to define a matrix for any symmetry operation. For example, the centre of symmetry converts a set of coordinates (x, y, z) into $(-x, -y, -z)$ and could be represented by the matrix

$$\begin{pmatrix} -1 & 0 & 0 \\ 0 & -1 & 0 \\ 0 & 0 & -1 \end{pmatrix}$$

(a) $\begin{pmatrix} 0 & 1 & 0 & 0 \\ 0 & 0 & 1 & 0 \\ 1 & 0 & 0 & 0 \\ 0 & 0 & 0 & 1 \end{pmatrix}$ (b) $\begin{pmatrix} 0 & 0 & 0 & 1 \\ 1 & 0 & 0 & 0 \\ 0 & 1 & 0 & 0 \\ 0 & 0 & 1 & 0 \end{pmatrix}$

(c) $\begin{pmatrix} 0 & 0 & 1 & 0 \\ 0 & 0 & 0 & 1 \\ 0 & 1 & 0 & 0 \\ 1 & 0 & 0 & 0 \end{pmatrix}$ (d) $\begin{pmatrix} 0 & 0 & 1 & 0 \\ 0 & 1 & 0 & 0 \\ 0 & 0 & 0 & 1 \\ 1 & 0 & 0 & 0 \end{pmatrix}$

Chemical background

Methane is said to belong to the point group known as T_d. The point group characterizes the symmetry elements possessed by a molecule; in the case of T_d, these are listed as E, $8C_3$, $3C_2$, $6S_4$, $6\sigma_d$. E is an identity operation which leaves the molecule unchanged. C_3 represents a rotation of $360°/3$ (i.e. $120°$), and C_2 a rotation of $360°/2$ (i.e. $180°$). S_4 is known as an improper rotation and involves rotation of $360°/4$ (i.e. $90°$) followed by reflection across the symmetry plane perpendicular to the axis of rotation. σ_d is a dihedral symmetry plane which bisects the angle formed by a pair of C_2 axes.

Solution to worked example

Application of the first symmetry operator matrix requires us to perform the multiplication

$$(H_A \quad H_B \quad H_C \quad H_d) \begin{pmatrix} 0 & 1 & 0 & 0 \\ 0 & 0 & 1 & 0 \\ 1 & 0 & 0 & 0 \\ 0 & 0 & 0 & 1 \end{pmatrix}$$

This is a 1×4 matrix multiplied by a 4×4 matrix, and will produce a 1×4 matrix result.

We calculate this as shown in the example above, but the presence of several zeros in the symmetry operator matrix assists this process considerably. We will generate the 1×4 matrix by matching elements in the initial 1×4 matrix with

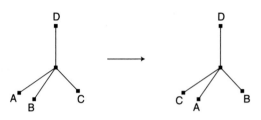

Figure 5.10 Symmetry operations on a methane molecule.

elements in each successive column of the 4×4 matrix. For the first column, we obtain

$$(H_A \times 0) + (H_B \times 0) + (H_C \times 1) + (H_D \times 0) = H_C$$

Subsequent columns give

$$(H_A \times 1) + (H_B \times 0) + (H_C \times 0) + (H_D \times 0) = H_A$$

$$(H_A \times 0) + (H_B \times 1) + (H_C \times 0) + (H_D \times 0) = H_B$$

$$(H_A \times 0) + (H_B \times 0) + (H_C \times 0) + (H_D \times 1) = H_D$$

so that the final matrix is

$$(H_C \quad H_A \quad H_B \quad H_D)$$

Comparison with the initial matrix shows that H_A has moved to H_B, H_B to H_C, H_C to H_A and H_D has remained in the same position. The matrix consequently represents a rotation of $120°$ about the $C–H_D$ bond, and the net result of performing this symmetry operation is shown in Figure 5.10.

Similar calculations give the following results for applications of the remaining symmetry operator matrices:

(b) $(H_B \quad H_C \quad H_D \quad H_A)$
(c) $(H_D \quad H_C \quad H_A \quad H_B)$
(d) $(H_D \quad H_B \quad H_A \quad H_C)$

Exercises

1. In the right-angled triangle shown below, calculate $\sin \theta$, $\cos \theta$ and $\tan \theta$, giving your answers correct to 2 d.p.

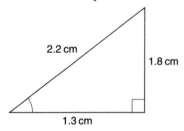

2. Calculate these, giving your answers correct to 3 d.p.
   (a) $\sin 11°$
   (b) $\cos(-32°)$
   (c) $\tan(\pi/8)$
3. Solve the equations, giving your answer in degrees.
   (a) $\sin x = 0.352$
   (b) $\cos x = -0.108$
4. Solve the equations, giving your answer in radians.
   (a) $\tan(2x + 1) = 0.472$
   (b) $\cos(3x + 2) = 0.564$
5. Calculate the modulus of these vectors.
   (a) $2\mathbf{i} - 2\mathbf{j} - 3\mathbf{k}$
   (b) $2\mathbf{i} + 4\mathbf{j} - 6\mathbf{k}$
6. If $\mathbf{a} = \mathbf{i} + 2\mathbf{j} - 3\mathbf{k}$ and $\mathbf{b} = 2\mathbf{i} - \mathbf{j} + \mathbf{k}$, calculate
   (a) $\mathbf{a} + \mathbf{b}$
   (b) $\mathbf{a} - \mathbf{b}$
   (c) $\mathbf{a} \cdot \mathbf{b}$
   (d) $\mathbf{a} \times \mathbf{b}$
7. Determine the angle between the vectors $2\mathbf{i} + 2\mathbf{j} + \mathbf{k}$ and $\mathbf{i} - \mathbf{j} - 2\mathbf{k}$.
8. If

$$A = \begin{pmatrix} 2 & 2 \\ 3 & 1 \end{pmatrix} \quad \text{and} \quad B = \begin{pmatrix} 1 & 0 \\ 4 & 2 \end{pmatrix}$$

   calculate
   (a) $A + B$
   (b) $A - B$
   (c) AB
9. Calculate the product

$$\begin{pmatrix} 3 & 0 & 1 \\ 2 & 4 & 2 \\ 1 & 1 & 3 \end{pmatrix} \begin{pmatrix} 2 \\ 1 \\ 1 \end{pmatrix}$$

10. If

$$A = \begin{pmatrix} 1 & 2 \\ 4 & 2 \end{pmatrix} \quad \text{and} \quad B = \begin{pmatrix} 2 & 3 \\ 3 & 1 \end{pmatrix}$$

   calculate $(2A + B)A$.

Problems

1. The second-order reflections from a potassium chloride crystal occur at an angle of 29.2° when radiation of wavelength 1.537 Å is used. Calculate the spacing of the planes which are responsible for this diffraction.

2. Suppose that the atoms A and B in Figure 5.9 are not equivalent, so that the bond lengths AC and BC are different and

$$\mathbf{AC} = 0.5d\mathbf{i} + 0.5d\mathbf{j} + 0.4d\mathbf{k}$$

$$\mathbf{BC} = -0.5d\mathbf{i} - 0.5d\mathbf{j} + 0.6d\mathbf{k}$$

Calculate the new value of the bond angle.

3. The atoms in a body-centred cubic lattice are situated at the corners of a cube and at its centre, as shown below. What are the four unique interatomic vectors in this lattice, if the cube has side a?

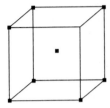

4. Calculate the volume of the unit cell in the mineral Malachite which has the formula $Cu_2(OH)_2CO_3$ and crystallizes in the space group $P2_1/a$. The unit cell dimensions are $a = 9.502$, $b = 11.974$, $c = 3.240$ Å and $\beta = 98.75°$.

5. The anti-cancer agent cisplatin has the structure as shown below, which also shows a set of coordinate axes which can be used to define the positions of each atom.

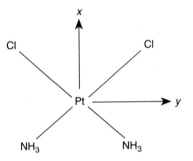

Write a matrix equation for finding the 2×1 matrix

$$\begin{pmatrix} x_2 \\ y_2 \end{pmatrix}$$

which denotes the position of N_2 or Cl_2 starting from the corresponding matrix containing x_1 and y_1.

Quantum mechanics 6

The ideas involved in quantum mechanics can be expressed quite concisely, yet a detailed study of individual quantum mechanical systems seems to rely heavily upon mathematics. In some ways this is unfortunate, as there is a danger that the underlying concepts may be lost in this detail. However, it is essential to be able to apply these ideas if we are to obtain results which are of use and which may be compared with experimental data. In this chapter, we will deal with the mathematics needed for some of the simpler concepts before looking at more complicated situations which need more specialized tools.

The concept of a photon is central to an appreciation of quantum mechanics. It is simply one unit of electromagnetic radiation, also known as a quantum. This is an example of the idea that quantities traditionally regarded as waves, in this case radiation, can behave as discrete particles.

6.1.1   Mathematical relationships

We saw earlier, in section 2.5.3, that if we know that a pair of variables are related proportionally then it is possible to obtain an expression for the relationship between them. Sometimes, we do not know what this relationship is, or even whether it exists at all. We then need to inspect the data given to see if it is possible to obtain such a relationship.

Worked example 6.1

Derive an equation which relates the energy E of a photon to its frequency v, given the following data:

$\dfrac{E}{10^{-19}\,\text{J}}$	$\dfrac{v}{10^{14}\,\text{s}^{-1}}$
19.89	30
9.945	15
6.630	10
4.973	7.5
3.978	6.0

Calculate the energy corresponding to a frequency of $8.57 \times 10^{14}\,\text{s}^{-1}$.

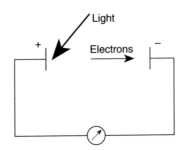

Figure 6.1　Apparatus used to demonstrate the photoelectric effect.

Chemical background

This is the basis of the well known photoelectric effect, which is often used as an introduction to quantum mechanics. The experimental arrangement for this demonstration is shown in Figure 6.1. The maximum kinetic energy of the electrons is proportional to the difference between the frequency of the incident radiation and the value known as the threshold frequency.

When light is absorbed by the surface of a metal, electrons are ejected, but this only happens when the frequency of light is above a certain value. Classically, we would expect light of high intensity but low frequency to also cause the ejection of electrons, and this discrepancy led Einstein to propose that light consisted of discrete photons.

Solution to worked example

Notice that the table headings are given as quotients including the units, as we have seen previously. Although it would be possible to expand each value in the table in terms of its units, it is actually neater *not* to do this but to work with the table headings as they are given.

A casual inspection of the data shows that as the values of E increase so do the values of v, as shown in Figure 6.2. The simplest relationship we could hope

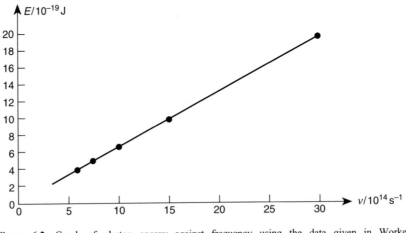

Figure 6.2　Graph of photon energy against frequency using the data given in Worked Example 6.1.

to find, then, would be that of direct proportion. In that case, the ratio E/v would be constant, and equal to the constant of proportionality. If the ratio E/v is constant, it is also true that the ratio

$$\frac{E/10^{-19}\,\text{J}}{v/10^{14}\,\text{s}^{-1}}$$

is a constant, this being easier to calculate from the data given. Our table of data can then be extended to include this calculation.

$\dfrac{E}{10^{-19}\,\text{J}}$	$\dfrac{v}{10^{14}\,\text{s}^{-1}}$	$\dfrac{E/10^{-19}\,\text{J}}{v/10^{14}\,\text{s}^{-1}}$
19.89	30	0.663
9.945	15	0.663
6.630	10	0.663
4.973	7.5	0.663
3.978	6.0	0.663

The constant ratio term indicates that E and v are indeed directly proportional and are related by the equation

$$\frac{E/10^{-19}\,\text{J}}{v/10^{14}\,\text{s}^{-1}} = 0.663$$

We can multiply both sides by $v/10^{14}\,\text{s}^{-1}$ so that this becomes

$$\frac{E}{10^{-19}\,\text{J}} = \frac{0.663 \times v}{10^{14}\,\text{s}^{-1}}$$

We can also multiply both sides by $10^{-19}\,\text{J}$ to obtain

$$E = 0.663 \times \frac{v}{10^{14}\,\text{s}^{-1}} \times 10^{-19}\,\text{J}$$

The powers of 10, $10^{-19}/10^{14}$ combine to give $10^{(-19-14)}$ or 10^{-33}, and the unit J/s^{-1} is the same as J s. This then gives the relationship as

$$E = (0.663 \times 10^{-33}\,\text{J s})v$$
$$= (6.63 \times 10^{-34}\,\text{J s})v$$

The quantity in brackets is known as **Planck's constant** and is normally given the symbol h, so we obtain the relationship

$$E = hv$$

where $h = 6.63 \times 10^{-34}\,\text{J s}$.

We can now substitute in the value of v of $8.57 \times 10^{14}\,\text{s}^{-1}$ to give

$$E = 6.63 \times 10^{-34}\,\text{J s} \times 8.57 \times 10^{14}\,\text{s}^{-1}$$
$$\approx 5.68 \times 10^{-19}\,\text{J}$$

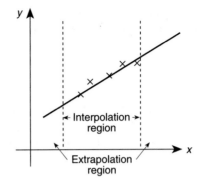

Figure 6.3   Interpolation and extrapolation.

since the units s and s^{-1} cancel. Notice that since the frequency value of $8.57 \times 10^{14} \, s^{-1}$ falls within the range of the values given in the question $(6.0–30 \times 10^{14} \, s^{-1})$ this is known as interpolation. If we had been calculating the energy by using a frequency value outside this range, it would be known as extrapolation (Figure 6.3). We can be less confident of extrapolating reliably since the relationship we have obtained may not be valid over other ranges of frequency.

6.2
Forces between atoms

Despite the successes of quantum mechanics, the classical view of atoms can still be very useful and is more appropriate in some circumstances. We are often interested in the interactions between atoms, as these allow the prediction of properties such as the energy and structure of a system.

6.2.1   Proportion

The subject of proportion was discussed in section 2.5.3.

Worked example 6.2

One model of a chemical bond is that of a simple spring, as shown in Figure 6.4, in which the force acting on the spring is proportional to the displacement of the bond length from its equilibrium value. In this case, the constant of proportionality is known as the force constant of the bond and has the value of $440 \, N \, m^{-1}$

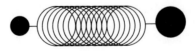

Figure 6.4   Displacement of a spring from its equilibrium position.

for a single bond between a pair of carbon atoms. Calculate the force on this bond if it is displaced by 2.5 pm from its equilibrium position.

Chemical background

Such calculations form the basis of the molecular mechanics method for calculating molecular structures and energies. The energy E arising from the displacement of a particular bond is calculated from an equation of the form

$$E = \tfrac{1}{2}k(l - l_0)^2$$

where k is the force constant, l the observed length and l_0 the equilibrium length. Figure 6.5 shows a plot of E against l for the displacement of the bond being discussed here.

Typical values of these parameters for a single bond between a pair of carbon atoms would be $k = 4.40$ mdyn Å^{-1} and $l_0 = 1.523$ Å. These non-SI units are invariably used in calculations of this type.

Solution to worked example

We need to set up an equation to relate the force F to the displacement x. Since $F \propto x$, it follows that

$$F = kx$$

and we are given the value of the proportionality constant k in the question. We substitute the values given into this equation to obtain

$$F = 440 \text{ N m}^{-1} \times 2.5 \text{ pm}$$
$$= 440 \text{ N m}^{-1} \times 2.5 \times 10^{-12} \text{ m}$$
$$= 1.1 \times 10^{-9} \text{ N}$$
$$= 1.1 \text{ nN}$$

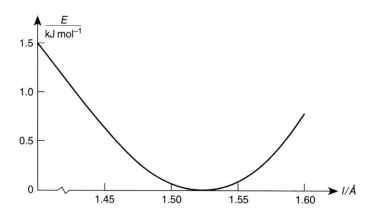

Figure 6.5   The energy arising from the displacement of a bond from its equilibrium length.

6.2.2   Stationary points

Stationary points were discussed in section 3.6.2.

Worked example 6.3

The constants A and B are related to the depth of the potential well ε and the equilibrium intermolecular separation r_e. For xenon these quantities have the values 1.9 kJ mol^{-1} and 4.06 Å respectively, the potential energy curve being illustrated in Figure 6.6.

The energy $E(r)$ of the interaction between two molecules separated by a distance r is given by the function

$$E(r) = \frac{-A}{r^6} + \frac{B}{r^{12}}$$

Locate and identify the stationary points of the function $E(r)$.

Chemical background

The function given represents the interaction energy between a pair of molecules, but such functions have been used as 'interatomic potentials' to model the interactions of atoms in the solid phase. Other types of function can be used to represent these interactions, and parameters such as A and B need to be varied to give the best agreement with experimental data.

Solution to worked example

We saw in section 3.6.2 that the stationary points of a function $f(x)$ were given by those values of x which make

$$\frac{df(x)}{dx} = 0$$

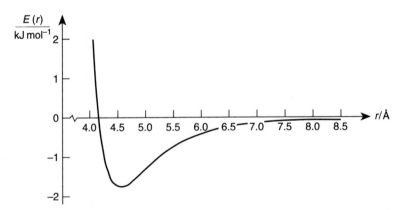

Figure 6.6   Graph of the interaction energy $E(r)$ as defined in Worked Example 6.3.

and that the nature of those stationary points is determined by evaluating the sign of

$$\frac{d^2 f(x)}{dx^2}$$

The differentiation of $E(r)$ is easier to perform if we rewrite it as

$$E(r) = -Ar^{-6} + Br^{-12}$$

since

$$\frac{1}{x^n} = x^{-n}$$

Using the rule that

$$\frac{d(ax^n)}{dx} = anx^{n-1}$$

then gives us

$$\frac{d(-Ar^{-6})}{dr} = (-A)(-6)r^{-6-1}$$

$$= 6Ar^{-7}$$

and

$$\frac{d(Br^{-12})}{dr} = B \times -12r^{-12-1}$$

$$= -12Br^{-13}$$

so that

$$\frac{dE(r)}{dr} = \frac{6A}{r^7} - \frac{12B}{r^{13}}$$

This will be zero when

$$\frac{6A}{r^7} = \frac{12B}{r^{13}}$$

This equation can be rearranged by multiplying both sides by r^{13} and dividing both sides by $6A$ to give

$$\frac{r^{13}}{r^7} = \frac{12B}{6A}$$

$$r^{13-7} = \frac{2B}{A}$$

$$r^6 = \frac{2B}{A}$$

We can finally solve this equation by taking the sixth root of either side, since

$$\sqrt[6]{r^6} = r$$

In other words, the process of taking the sixth root is the inverse of raising a number to power 6. The same relationship holds for any other power. Having done this, we finally obtain

$$r = \sqrt[6]{\frac{2B}{A}}$$

This tells us the position of the stationary point but nothing about its nature. We now need to calculate $d^2E(r)/dr^2$ which we can do by differentiating the equation

$$\frac{dE(r)}{dr} = \frac{6A}{r^7} - \frac{12B}{r^{13}}$$

which is easier to do if we rewrite it as

$$\frac{dE(r)}{dr} = 6Ar^{-7} - 12Br^{-13}$$

Using the same rules of differentiation as before gives

$$\frac{d^2E}{dr^2} = (6 \times -7)Ar^{-7-1} + (-12 \times -13)Br^{-13-1}$$

$$= -42Ar^{-8} + 156Br^{-14}$$

$$= -\frac{42A}{r^8} + \frac{156B}{r^{14}}$$

Since the minimum occurs when

$$r = \sqrt[6]{\frac{2B}{A}}$$

it follows that $r^6 = 2B/A$.

It helps if we can rewrite our expression for d^2E/dr in terms of powers of r^6.

$$\frac{d^2E}{dr} = \frac{-42A}{r^8} + \frac{156B}{r^{14}}$$

$$= \frac{1}{r^2}\left(-\frac{42A}{r^6} + \frac{156B}{r^{12}}\right)$$

Now,

$$r^{12} = (r^6)^2$$

$$= \left(\frac{2B}{A}\right)^2$$

$$= \frac{2^2 B^2}{A^2}$$

$$= \frac{4B^2}{A^2}$$

and so

$$\frac{d^2 E}{dr^2} = \frac{1}{r^2}\left(-\frac{42A}{2B/A} + \frac{156B}{4B^2/A^2}\right)$$

$$= \frac{1}{r^2}\left(-\frac{42A^2}{2B} + \frac{156BA^2}{4B^2}\right)$$

$$= \frac{1}{r^2}\left(-\frac{21A^2}{B} + \frac{39A^2}{B}\right)$$

$$= \frac{1}{r^2}\frac{18A^2}{B}$$

Since A and B are both positive constants $d^2 E/dr^2$ must be greater than zero, indicating a minimum.

6.3
Particle in a box

One of the first quantum mechanical systems treated is usually the particle in a box. This consists of a particle constrained within a one-dimensional box of a specified width, within which the potential energy is zero and outside of which the potential energy is infinite. The model is readily extended to two- and three-dimensional systems.

6.3.1   Complex numbers

We met the concept of complex numbers in section 5.3.1. In particular, we saw that a complex exponential could be expressed by the equation

$$e^{ikx} = \cos kx + i \sin kx$$

Worked example 6.4

The wavefunctions ψ for the particle in a one-dimensional box are given by the expression

$$\psi = \left(\frac{2}{L}\right)^{1/2}\left(\frac{1}{2i}\right)(e^{ikx} - e^{-ikx})$$

where $i^2 = -1$, L is the length of the box, x is the position of the particle within the box and k is a constant. Simplify this expression.

Chemical background

The constant k is actually given by the formula

$$k = \frac{\sqrt{8\pi^2 mE}}{h}$$

where m is the mass of the particle, E its energy and h is Planck's constant. If we apply the boundary conditions that the wavefunction must fall to zero at $x = 0$ and $x = L$, we can solve this expression to give the allowed values of the energy, as we will see in the next worked example.

Solution to worked example

We need to use the relationships

$$e^{ikx} = \cos kx + i \sin kx$$

and

$$e^{-ikx} = \cos kx - i \sin kx$$

We then have

$$\psi = \left(\frac{2}{L}\right)^{1/2}\left(\frac{1}{2i}\right)(e^{ikx} - e^{-ikx})$$

$$= \left(\frac{2}{L}\right)^{1/2}\left(\frac{1}{2i}\right)[\cos kx + i \sin kx - (\cos kx - i \sin kx)]$$

$$= \left(\frac{2}{L}\right)^{1/2}\left(\frac{1}{2i}\right)[\cos kx + i \sin kx - \cos kx + i \sin kx]$$

The two terms in $\cos kx$ cancel out and we are left with

$$\psi = \left(\frac{2}{L}\right)^{1/2}\left(\frac{1}{2i}\right)[2i \sin kx]$$

$$= \left(\frac{2}{L}\right)^{1/2} \sin kx$$

The graph of this wavefunction is shown in Figure 6.7.

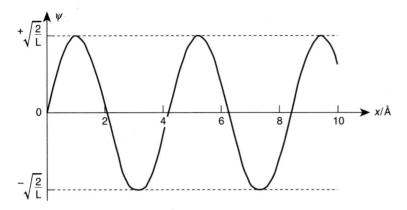

Figure 6.7   Graph of the wavefunction for the particle in a one-dimensional box.

6.3.2   Sequences

In section 2.2.3 we met the idea of a function for the first time. This was represented by $f(x)$, where x was simply a number upon which certain operations (such as multiplication by a constant, raising to a power and so on) were performed. A sequence differs from a function only in that x is replaced by n, where n is an integer (i.e. a whole number) and is therefore restricted to the values 0, 1, 2, 3, 4, ...

For example, we could define a sequence as

$$f(n) = 3n + 1 \quad \text{for } n \geq 0$$

We would then have

$$f(0) = (3 \times 0) + 1 = 1$$
$$f(1) = (3 \times 1) + 1 = 4$$
$$f(2) = (3 \times 2) + 1 = 7$$
$$f(3) = (3 \times 3) + 1 = 10$$

and so on. It is necessary to specify the starting value of n, which will usually be 0 or 1. This can be done using a greater than sign $>$ or a greater than or equal to sign $\geq$.

In quantum mechanics, we are often concerned with the idea that particles may only occupy discrete energy levels which are also labelled as integers (usually from 1 upwards).

Worked example 6.5

The energy E of each level n in the one-dimensional particle in a box problem is given by the function

$$E(n) = \frac{n^2 h^2}{8ma^2}$$

for $n \geq 1$ where m is the mass of the particle, a is the width of the box and h is Planck's constant. Find the sum of the first five terms of this sequence in terms of a, h and m.

Chemical background

The pi electrons in hexatriene are delocalized, and each can be thought of as being contained in a one-dimensional box of length 7.3 Å, this being the distance between the terminal carbon atoms, plus half a bond length on either side.

Solution to worked example

We clearly need to add together five terms of the form

$$\frac{n^2h^2}{8ma^2}$$

Once we realize that the quantity $h^2/8ma^2$ is common to each of these terms, we can write the required sum simply as

$$\frac{h^2}{8ma^2}(1^2 + 2^2 + 3^2 + 4^2 + 5^2) = \frac{h^2}{8ma^2}(1 + 4 + 9 + 16 + 25)$$

$$= \frac{55h^2}{8ma^2}$$

These energy levels are shown in Figure 6.8.

This is analogous to summing the energies of electrons in a molecule where each occupies a separate energy level.

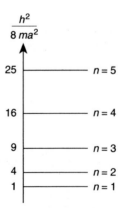

Figure 6.8   The energy levels available to a particle confined to a one-dimensional box.

6.3.3 Inverse functions

This topic has been dealt with in section 4.6.2.

Worked example 6.6

The quantum mechanical energy E of a particle in a three-dimensional box is given by the function

$$E(n_x, n_y, n_z) = \frac{h^2}{8m}\left(\frac{n_x^2}{a^2} + \frac{n_y^2}{b^2} + \frac{n_z^2}{c^2}\right)$$

where h is Planck's constant and m the mass of the particle. The quantum numbers n_x, n_y and n_z relate to the directions x, y and z in which the box has dimensions a, b and c respectively.

Obtain an expression for the inverse function arc $E(n)$ in the special case where $n_x = n_y = n_z = n$ and $a = b = c = d$ which is the situation existing in a cube.

Since we are now dealing with three quantum numbers, it is possible to obtain degenerate energy levels. These have the same energies but different quantum numbers, for example $E(1, 2, 1)$ and $E(1, 1, 2)$ when $a = b = c$.

Chemical background

The extension of the quantum mechanics of the particle in a one-dimensional box to three dimensions is relatively straightforward. This is due to the fact that the motion in the three perpendicular directions x, y and z can be treated independently. If this was not so, the solution to the problem would be much more difficult to obtain. This ability to partition the energy has important implications in both quantum mechanics and statistical mechanics.

This model could be used to represent, say, nitrogen gas in a 10 cm by 10 cm by 10 cm container. The calculation would then show that the separation between the energy levels is extremely small, and so there is no effective quantization on this scale.

Solution to worked example

The first stage in this problem is to obtain an expression for $E(n)$, the energy of a particle in a cube. If we make the substitutions

$$n_x = n_y = n_z = n$$

and

$$a = b = c = d$$

into the given equation, we obtain

$$E(n, n, n) = \frac{h^2}{8m}\left(\frac{n^2}{d^2} + \frac{n^2}{d^2} + \frac{n^2}{d^2}\right)$$

$$= \frac{3h^2n^2}{8md^2}$$

Before we consider the inverse of this function, arc $E(n)$, it is worth considering the stages involved in calculating $E(n)$ from a given value of n:

1. Square n to give n^2.
2. Multiply n^2 by $3h^2$ to give $3h^2n^2$.
3. Divide $3h^2n^2$ by $8md^2$ to give

$$\frac{3h^2n^2}{8md^2}$$

The reverse of this process is then as follows:

1. Multiply n by $8md^2$ to give $8md^2n$.
2. Divide $8md^2n$ by $3h^2$ to give $8md^2n/3h^2$.
3. Take the square root of $8ma^2n/3h^2$ to give

$$\text{arc } E(n) = \sqrt{\frac{8md^2n}{3h^2}}$$

6.4
The free particle

Although a free particle which is not subject to any forces is not a particularly interesting system from a chemical point of view, it does allow us to work through a quantum mechanical treatment at a relatively simple level. A knowledge of the potential energy of the system allows the Schrödinger equation to be written and subsequently solved to give the general wavefunction. We then need to apply the boundary conditions and normalize the function before it can be used to predict properties.

6.4.1   The complex conjugate

We saw in section 5.3.1 that a complex number z was defined in terms of real numbers a and b as

$$z = a + ib$$

where $i^2 = -1$. The complex conjugate of z, known as z^*, is defined as

$$z^* = a - ib$$

So, to obtain the conjugate of a complex number, we change the sign of the imaginary part. If $z = 2 + 3i$, we would then have $z^* = 2 - 3i$. Similarly, if $z = 2 - 4i$, we would have $z^* = 2 + 4i$. z and z^* are called a **complex conjugate pair**; each one is the complex conjugate of the other.

Notice that since

$$e^{ikx} = \cos kx + i \sin x$$

$$e^{-ikx} = \cos kx - i \sin x$$

e^{ikx} and e^{-ikx} are a complex conjugate pair.

6.4.2 The modulus of a complex number

We saw in the previous chapter that the modulus of a vector was its length, without regard to direction. Complex numbers can be represented in a similar fashion to vectors by use of an Argand diagram. This plots the real part a of a complex number in the x-direction and the imaginary part b in the y-direction. Such a plot for the complex number $1 + 4i$ is shown in Figure 6.9.

The modulus of the complex number z is then given as

$$|z| = \sqrt{a^2 + b^2}$$

This can be compared with the expression for calculating the modulus of a vector which we met in section 5.2.3. If we now consider the product of the complex conjugate of z with z, we get

$$z^*z = (a - ib)(a + ib)$$
$$= a(a + ib) - ib(a + ib)$$
$$= a^2 + aib - iba - i^2b^2$$

Since $aib = iba$ and $i^2 = -1$ we obtain

$$z^*z = a^2 + b^2$$

and taking the square root of both sides of this equation gives

$$\sqrt{z^*z} = \sqrt{a^2 + b^2} = |z|$$

so that the modulus of a complex number can be obtained by taking the square root of the product of the complex number and its complex conjugate.

Worked example 6.7

One wavefunction ψ which satisfies the Schrödinger equation for a free particle which moves only in the x-direction is given by the equation

$$\psi = A \exp\left(-i\sqrt{8\pi^2 mE}\,\frac{x}{h}\right)$$

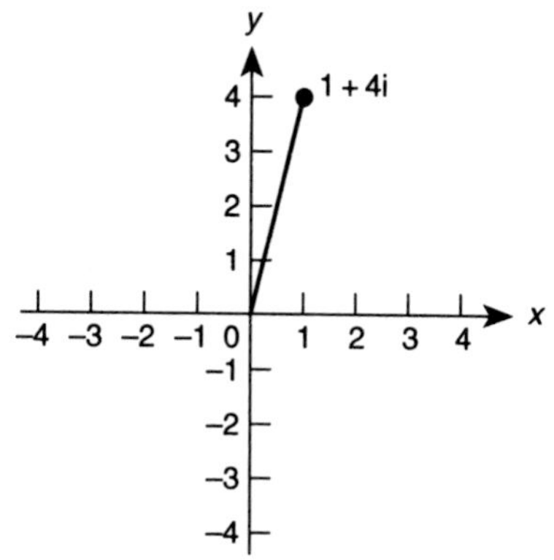

Figure 6.9 Argand diagram for $1 + 4i$.

where A is a constant, h is Planck's constant, m is the mass of the particle and E its energy when at position x. The probability of finding the particle at a particular position is given by the function $\psi^*\psi$. Determine the equation which describes this function.

Chemical background

An example of a free particle in chemistry is an ionized electron, which may have any energy value. The Balmer series in the spectrum of hydrogen converges when such electrons fall to the energy level with principal quantum number 2.

Since the final value for the probability does not depend on x, the free particle has an equal probability of being found anywhere in the x-direction, and it is said to be nonlocalized.

Solution to worked example

From the expression for the wavefunction

$$\psi = A \exp\left(-i\sqrt{8\pi^2 mE}\,\frac{x}{h}\right)$$

we can immediately write down an expression for the conjugate wavefunction

$$\psi^* = A^* \exp\left(i\sqrt{8\pi^2 mE}\,\frac{x}{h}\right)$$

where A^* is the complex conjugate of A. We are now able to form the product of these two functions:

$$\psi^*\psi = A^* \exp\left(i\sqrt{8\pi^2 mE}\,\frac{x}{h}\right) A \exp\left(-i\sqrt{8\pi^2 mE}\,\frac{x}{h}\right)$$

and since $e^{-x} = 1/e^x$ this can be rewritten as

$$\psi^*\psi = A^*A\,\frac{\exp(i\sqrt{8\pi^2 mE}\,x/h)}{\exp(i\sqrt{8\pi^2 mE}\,x/h)}$$

$$= A^*A$$

$$= |A|$$

6.5
The hydrogen atom wavefunction

In its ground state, the single electron in a hydrogen atom occupies the 1s shell; this could be promoted to the 2s or 2p level in the excited state. The numbers 1 and 2 are known as the principal quantum number.

Wavefunctions can be defined for the electron in a hydrogen atom in both its ground state and in higher energy states. Their form is found by solving the appropriate Schrödinger equation, and details of this procedure can be found in physical chemistry textbooks. The wavefunctions are normally expressed as functions of three variables, which are the distance r of the electron from the nucleus, the zenith angle θ and the azimuthal angle ϕ, as shown in Figure 6.10. These quantities are known as spherical polar coordinates and define the position of a particle in space, and are more useful than the more familiar

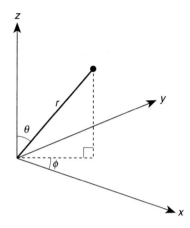

Figure 6.10   The definition of spherical polar coordinates.

cartesian coordinates x, y and z when we are dealing with systems such as this where the symmetry is spherical.

6.5.1   *Differentiation of a product and integration by parts*

We saw in section 2.3.1 how to differentiate simple functions. We also come across *products* of functions which need to be differentiated, such as $x^2 e^x$. In this case, we need to be able to apply the standard rule for the differentiation of a product.

If our function $f(x)$ can be expressed as the product of two functions $u(x)$ and $v(x)$, such that

$$f(x) = u(x)v(x)$$

then

$$\frac{\mathrm{d}f(x)}{\mathrm{d}x} = v(x)\frac{\mathrm{d}u(x)}{\mathrm{d}x} + u(x)\frac{\mathrm{d}v(x)}{\mathrm{d}x}$$

This is perhaps more easily remembered in words as 'differentiate the first and multiply by the second, differentiate the second and multiply by the first'.

In our example above, we then have

$$u(x) = x^2 \qquad v(x) = e^x$$

and

$$\frac{\mathrm{d}u(x)}{\mathrm{d}x} = 2x \qquad \frac{\mathrm{d}v(x)}{\mathrm{d}x} = e^x$$

so

$$\frac{d(x^2 e^x)}{dx} = e^x \times 2x + x^2 \times e^x$$

$$= 2x\, e^x + x^2\, e^x$$

The equation expressing the rule for differentiating a product can also be of use to us when integrating certain more complicated functions.

$$\frac{d(uv)}{dx} = v\frac{du}{dx} + u\frac{dv}{dx}$$

can be rearranged as

$$u\frac{dv}{dx} = \frac{d(uv)}{dx} - v\frac{du}{dx}$$

We can now integrate this expression.

$$\int u\frac{dv}{dx}\, dx = \int \frac{d(uv)}{dx}\, dx - \int v\frac{du}{dx}\, dx$$

Since integration is the reverse of differentiation, integrating a derivative gives back the original function and so we have

$$\int \frac{d(uv)}{dx}\, dx = uv$$

which gives us

$$\int u\frac{dv}{dx}\, dx = uv - \int v\frac{du}{dx}\, dx$$

Here u is a function that you will need to differentiate, and dv/dx is a function you will need to integrate. This equation is used to perform the integration process known as **integration by parts**. Now consider the calculation of the integral

$$\int x \ln x\, dx$$

we can integrate and differentiate the term x, but with $\ln x$ we would prefer to differentiate. So, if we make the substitutions

$$u = \ln x \quad \text{and} \quad \frac{dv}{dx} = x$$

this gives, on differentiating u and integrating dv/dx

$$\frac{du}{dx} = \frac{1}{x} \quad \text{and} \quad v = \frac{x^2}{2}$$

ignoring for the time being, the constant of integration. We can now substitute into the general equation

$$\int u \frac{dv}{dx} dx = uv - \int v \frac{du}{dx} dx$$

to give

$$\int x \ln x \, dx = \frac{x^2}{2} \ln x - \int \frac{x^2}{2} \frac{1}{x} dx$$

$$= \frac{x^2}{2} \ln x - \int \frac{x}{2} dx$$

$$= \frac{x^2}{2} \ln x - \frac{x^2}{4} + C$$

where C is the constant of integration, normally inserted at the end of the calculation for simplicity. It is easy to forget it!

6.5.2   Calculus of the exponential function

The exponential function e^x is the only function which does not change upon differentiation and integration. We now need to use the slightly more general result for the function e^{ax} where a is a constant. Examples are e^{3x} or e^{-2x}. The rules we need are

$$\frac{d(e^{ax})}{dx} = a \, e^{ax} \quad \text{and} \quad \int e^{ax} \, dx = \frac{e^{ax}}{a} + C$$

Some examples of these are

$$\frac{d(e^{-x})}{dx} = -e^{-x}$$

$$\frac{d(e^{4x})}{dx} = 4e^{4x}$$

$$\int e^{-3x} \, dx = \frac{e^{-3x}}{-3} + C$$

Worked example 6.8

During the normalization of the wavefunction representing the lowest electronic state of the hydrogen atom, the integral

$$\int_0^\infty r^2 \exp\left(\frac{-2r}{a_0}\right) dr$$

needs to be evaluated. Use integration by parts to evaluate this integral.

Chemical background

The process of normalization is used to ensure that the probability of finding the specified particle within a given volume is correct. Since the electron must be found somewhere within the hydrogen atom, the probability of finding it is 1, and so the probability function derived from the wavefunction must integrate over all space to give this value. It can be adjusted by the inclusion of a normalization factor, which is usually given the symbol N.

Solution to worked example

The function to be integrated is shown in Figure 6.11. We are obviously going to use the equation given above for performing integration by parts. This problem is given in terms of the variable r, so the equation we need to use is actually

$$\int u \frac{dv}{dr}\, dr = uv - \int v \frac{du}{dr}\, dr$$

We now need to assign the two functions $u(r)$ and $v(r)$ to the two functions r^2 and $\exp(-2r/a_0)$. If we set $u(r) = r^2$, then we obtain, by differentiation

$$\frac{du}{dr} = 2r$$

whereas if we set $dv/dr = r^2$ we obtain, by integration

$$v(r) = \tfrac{1}{3}r^3$$

The first of these seems to be leading towards a simpler calculation, but we also need to consider what happens to the second function.

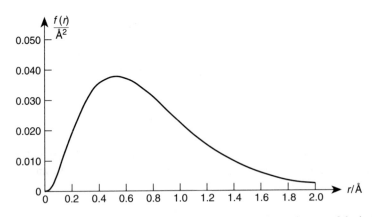

Figure 6.11 Graph of the wavefunction representing the lowest electronic state of the hydrogen atom.

If we have $u(r) = \exp(-2r/a_0)$ then, by differentiating, we obtain

$$\frac{du}{dr} = \frac{-2}{a_0} \exp\left(\frac{-2r}{a_0}\right)$$

whereas the substitution $dv/dr = \exp(-2r/a_0)$ gives, on integration,

$$v(r) = \frac{\exp\left(\dfrac{-2r}{a_0}\right)}{-2/a_0}$$

$$= -\frac{a_0}{2} \exp\left(\frac{-2r}{a_0}\right)$$

There is no particular advantage in either of these substitutions for the exponential term, so we choose (arbitrarily) to take

$$u(r) = r^2 \quad \text{and} \quad \frac{dv}{dr} = \exp\left(\frac{-2r}{a_0}\right)$$

so that

$$\frac{du}{dr} = 2r \quad \text{and} \quad v = -\frac{a_0}{2} \exp\left(\frac{-2r}{a_0}\right)$$

Making these substitutions into our general equation

$$\int u \frac{dv}{dr} \, dr = uv - \int v \frac{du}{dr} \, dr$$

gives

$$\int r^2 \exp\left(\frac{-2r}{a_0}\right) = r^2\left(-\frac{a_0}{2} \exp\left(\frac{-2r}{a_0}\right)\right) - \int \left(-\frac{a_0}{2} \exp\left(\frac{-2r}{a_0}\right)\right) 2r \, dr$$

$$= -\frac{a_0}{2} r^2 \exp\left(\frac{-2r}{a_0}\right) + a_0 \int r \exp\left(\frac{-2r}{a_0}\right) dr$$

At this point in a problem involving the use of integration by parts we would now evaluate the integral in the second term on the right of the equation which is

$$\int r \exp\left(\frac{-2r}{a_0}\right) dr$$

In this case, however, it is not at all obvious how this can be done. However, a closer inspection reveals that this integral can also be evaluated using integration by parts. If we set

$$u = r \quad \text{and} \quad \frac{dv}{dr} = \exp\left(\frac{-2r}{a_0}\right)$$

we have

$$\frac{du}{dr} = 1 \quad \text{and} \quad v(r) = \frac{\exp\left(\dfrac{-2r}{a_0}\right)}{-2/a_0} = -\frac{a_0}{2}\exp\left(\frac{-2r}{a_0}\right)$$

Substituting these values into

$$\int u\frac{dv}{dr} = uv - \int v\frac{du}{dr}\,dr$$

gives us

$$\int r\exp\left(\frac{-2r}{a_0}\right)dr = -\frac{a_0}{2}r\exp\left(\frac{-2r}{a_0}\right) - \int\left(-\frac{a_0}{2}\exp\left(\frac{-2r}{a_0}\right)\right)\times 1\,dr$$

$$= -\frac{a_0}{2}r\exp\left(\frac{-2r}{a_0}\right) + \frac{a_0}{2}\int \exp\left(\frac{-2r}{a_0}\right)dr$$

$$= -\frac{a_0}{2}r\exp\left(\frac{-2r}{a_0}\right) + \frac{a_0}{2}\frac{\exp\left(\dfrac{-2r}{a_0}\right)}{-2/a_0}$$

$$= -\frac{a_0}{2}r\exp\left(\frac{-2r}{a_0}\right) - \frac{a_0^2}{4}\exp\left(\frac{-2r}{a_0}\right)$$

Notice that we have been able to evaluate this integral completely, so this expression can now be substituted into our equation for the original integral and we obtain

$$\int r^2\exp\left(\frac{-2r}{a_0}\right)dr = -\frac{a_0}{2}r^2\exp\left(\frac{-2r}{a_0}\right) + a_0\left(-\frac{a_0}{2}r\exp\left(\frac{-2r}{a_0}\right) - \frac{a_0^2}{4}\exp\left(\frac{-2r}{a_0}\right)\right)$$

$$= -\frac{a_0}{2}r^2\exp\left(\frac{-2r}{a_0}\right) - \frac{a_0^2}{2}r\exp\left(\frac{-2r}{a_0}\right) - \frac{a_0^3}{4}\exp\left(\frac{-2r}{a_0}\right)$$

$$= \exp\left(\frac{-2r}{a_0}\right)\left(-\frac{a_0 r^2}{2} - \frac{a_0^2 r}{2} - \frac{a_0^3}{4}\right)$$

We have not yet included the limits on our integration sign. It is actually easier to include them only at the end, so from our previous working we have

$$\int_0^\infty r^2\exp\left(\frac{-2r}{a_0}\right)dr = \left[\exp\left(\frac{-2r}{a_0}\right)\left(-\frac{a_0 r^2}{2} - \frac{a_0^2 r}{2} - \frac{a_0^3}{4}\right)\right]_0^\infty$$

Putting in the upper limit of $r = \infty$ first, and realizing that

$$\exp\left(\frac{-2r}{a_0}\right) = \frac{1}{\exp\left(\dfrac{+2r}{a_0}\right)}$$

shows that the whole of this term will tend to zero for the upper limit, since we have the reciprocal of a very large number. Multiplying by zero, gives a value of zero for the whole expression as $r \to \infty$.

On the other hand, when $r = 0$

$$\exp\left(\frac{-2r}{a_0}\right) = e^0 = 1$$

since any number raised to the power zero is one. The first two terms in brackets are zero when $r = 0$, and so we are left with

$$\int_0^\infty r^2 \exp\left(\frac{-2r}{a_0}\right) dr = 0 - \left(-\frac{a_0^3}{4}\right)$$

$$= \frac{a_0^3}{4}$$

6.6
The helium atom

The quantum mechanics of the helium atom are rather more complicated than those of the hydrogen atom because we now have to account for the interactions between electrons, as well as that between an electron and the nucleus. These interactions are illustrated in Figure 6.12. In practice, we build a trial wave-function based on the information we have already obtained for the 1s orbital of hydrogen.

6.6.1   Stationary points

The method for determining the stationary points of a function has been presented in section 3.5.2.

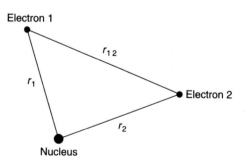

Figure 6.12   Interactions within the helium atom.

Worked example 6.9

The energy E of the helium atom can be expressed as

$$E = \left(\frac{-2B^2 m_e e^4}{h^2}\right)\left(-2(Z')^2 + \frac{27}{4}Z'\right)$$

where m_e is the mass of an electron, e is the electronic charge and h is Planck's constant. Z' is the effective nuclear charge on the helium atom and is treated as an adjustable parameter. Find the value of Z' which makes the value of E a minimum.

Chemical background

The method being used here is actually an example of the Variation Theorem, which finds a number of applications in quantum mechanics. It involves specifying a trial wavefunction and then optimizing a specified adjustable parameter.

This approach leads to a value for the energy of -77.5 eV which compares with an experimental value of -79.0 eV. A calculated value of -79.3 eV can be obtained by including a term for electron correlation in the 1s orbital trial wavefunction. The function for E used here is shown in Figure 6.13.

Solution to worked example

$$E = \left(\frac{-2B^2 m_e e^4}{h^2}\right)\left(-2(Z')^2 + \frac{27}{4}Z'\right)$$

We will obtain a minimum value of E when

$$\frac{dE}{dZ'} = 0 \quad \text{and} \quad \frac{d^2E}{dZ'^2} > 0$$

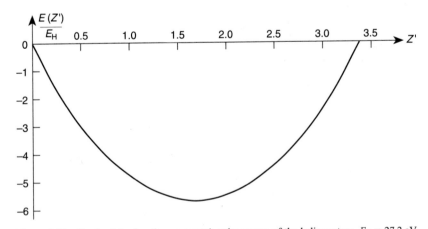

Figure 6.13 Graph of the function representing the energy of the helium atom. $E_H = 27.2$ eV.

The term in the first bracket of the expression for E is a constant and so this remains throughout the calculation. We know that

$$\frac{d(x^2)}{dx} = 2x \quad \text{and} \quad \frac{d(x)}{dx} = 1$$

so that

$$\frac{d}{dZ'}\left(-2(Z')^2 + \frac{27}{4}Z'\right) = -4Z' + \frac{27}{4}$$

This term will be zero when

$$-4Z' + \frac{27}{4} = 0$$

Adding $4Z'$ to both sides and dividing both sides by 4, isolates Z' to give

$$Z' = \frac{27}{16}$$

The full expression for the first derivative of E is

$$\frac{dE}{dZ'} = \left(\frac{-2B^2 m_e e^4}{h^2}\right)\left(-4Z' + \frac{27}{4}\right)$$

so the second derivative will be

$$\frac{d^2E}{dZ'^2} = -4\left(\frac{-2B^2 m_e e^4}{h^2}\right) = \frac{8B^2 m_e e^4}{h^2}$$

This is a positive value, and the second derivative being positive confirms that this stationary point is a minimum.

Exercises

1. From the following data deduce the relationship between the variables x and y:

x	1.42	3.62	4.75	6.24	7.08
y	7.12	2.79	2.13	1.62	1.43

2. Determine the location and the nature of the stationary points of the function $f(x) = 2x^2 - 4x + 8$.

   Hence state the location and nature of the stationary points of $g(x) = 2x^2 - 4x + 2$.

3. Calculate the product $z_1 z_2$ if

$$z_1 = 3\left(\cos\frac{\pi}{3} - i\sin\frac{\pi}{3}\right)$$

and

$$z_2 = 2\left(\cos\frac{\pi}{4} - i\sin\frac{\pi}{4}\right)$$

4. Find the next two terms of these sequences.
    (a) 4, 7, 10, 13, ...
    (b) 5, 10, 17, 26, ...
5. Find the general term of the sequence

$$3, 6, 12, 24, 48, \ldots$$

6. Find the inverse of the function $f(x) = 2e^{ax^2}$.
7. Differentiate these functions.
    (a) $f(x) = x^2 \ln x$
    (b) $g(x) = x\,e^{2x}$
8. (a) Differentiate the function $\exp(3x + 2)$.
    (b) Evaluate the integral

$$\int_0^3 e^{2x}\,dx$$

9. Use the method of integration by parts to evaluate this definite integral:

$$\int_0^1 x\,e^{2x}\,dx$$

10. By writing $\ln x$ as the product of 1 and $\ln x$, evaluate the integral

$$\int \ln x\,dx$$

using the method of integration by parts.

Problems

1. A function which can be used to describe the interaction energy $E(r)$ between a pair of molecules is

$$E(r) = \frac{A}{r^9} - \frac{B}{r^6}$$

where A and B are constants. Locate and identify the stationary points of this function.
2. The energy $E(n_x, n_y)$ of a particle in a two-dimensional box is given by the equation

$$E(n_x, n_y) = \frac{h^2}{8m}\left(\frac{n_x^2}{a^2} + \frac{n_y^2}{b^2}\right)$$

where n_x and n_y are quantum numbers in the x and y directions respectively, a and b are the dimensions of the box in the corresponding directions, h is

Planck's constant and m is the mass of the particle. Calculate the inverse function arc $E(n)$ in the special case when $n_x = n_y = n$ and $a = b = d$.

3. The following data were obtained for the maximum kinetic energy T_{max} of the electrons emitted from a sodium surface in a photoelectric experiment as a function of frequency v. Determine the relationship between T_{max} and v without using a graph.

T_{max}/eV	0.0	0.5	1.0	1.5	2.0	2.5
10^{-14} v/Hz	5.47	6.73	7.99	9.25	10.51	11.77

4. The wavefunction $\psi_{1s}(r)$ of the 1s orbital of the hydrogen atom is given by the equation

$$\psi_{1s}(r) = 2\left(\frac{Z}{a_0}\right)^{3/2} \exp\left(\frac{-Zr}{a_0}\right)$$

where Z is the nuclear charge, a_0 the Bohr radius and r the distance of the electron from the nucleus. Determine the inverse function arc $\psi_{1s}(r)$.

5. The Variation Theorem is often used to determine the wavefunctions of the helium atom. However, it can also be used with a trial wavefunction for the one-dimensional particle in a box. This involves the integral

$$\int_0^a (ax - x^2)\left(\frac{-\hbar^2}{2m}\frac{d^2}{dx^2}\right)(ax - x^2)\, dx$$

where a is the length of the box and x is the position of the particle of mass m within it. The constant $\hbar$ is $h/2\pi$ where h is Planck's constant. Perform the required differentiation of the function $ax - x^2$ and then evaluate this integral between the limits shown.

7 Spectroscopy

The subject of spectroscopy provides us with a description of how chemical particles interact with electromagnetic radiation. Spectroscopic techniques can be classified according to the type of incident radiation (such as ultraviolet or infra-red), or according to the molecular phenomena giving rise to the spectrum (such as rotation or vibration). Since we are concerned with interactions at the microscopic level, it is not surprising that the ideas of quantum mechanics are important in a development of this subject. In mathematical terms, most aspects of spectroscopy can be dealt with using techniques we have already met.

7.1 Electromagnetic radiation

Electromagnetic radiation is so called because it consists of an oscillating electric field and an oscillating magnetic field at right angles to one another, as shown in Figure 7.1. The electromagnetic spectrum ranges from gamma rays with a wavelength of about 10^{-12} m through to radio waves whose wavelength is approximately 100 m. In between, in ascending order of wavelength, are X-rays, ultraviolet radiation, visible light, infra-red radiation, microwaves and television waves (Figure 7.2). As well as being characterized by wavelength λ, electromagnetic radiation can be expressed in terms of its frequency ν.

7.1.1 Direct and inverse proportion

These types of proportion were discussed in section 2.5.3.

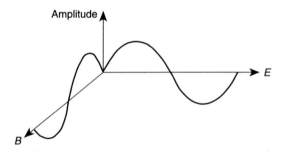

Figure 7.1 Electromagnetic radiation consisting of an electric field E and a magnetic field B.

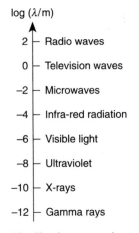

Figure 7.2   The electromagnetic spectrum.

Worked example 7.1

The relationship between the wavelength λ and frequency v of electromagnetic radiation is

$$c = \lambda v$$

where c is its velocity, which can be taken as a constant. How can the relationship between λ and v be described?

The velocity of light is exactly $299\,792\,458$ m s^{-1}. As well as being measured experimentally, this value can be predicted from Maxwell's equations which describe the behaviour of electric and magnetic fields.

Chemical background

The energy E of a photon is given by the equation

$$E = hv$$

where h is Planck's constant. Therefore, as the frequency of the radiation increases so does its energy. Since the wavelength is inversely proportional to the frequency, it also follows that higher wavelength radiation will have a lower energy.

An example is that of ultraviolet light which has a lower wavelength but higher energy and frequency than that of visible light. Consequently, it is more effective in inducing photochemical reactions.

Solution to worked example

Since the product of the two variables is a constant, we should be able to recognize instantly that they are inversely proportional. Alternatively, we can rearrange the equation given, by dividing both sides by λ, so that λ appears on one side and v on the other.

$$\frac{c}{\lambda} = v$$

It should now be apparent that as we increase λ then v must decrease.

Worked example 7.2

In addition to wavelength and frequency, a quantity often used in spectroscopy is wavenumber. This is normally given the symbol $\tilde{v}$, defined as the reciprocal of the wavelength, i.e.

$$\tilde{v} = \frac{1}{\lambda}$$

and has the units cm^{-1}. What is the wavenumber of the sodium D line having frequency 5.086×10^8 MHz? Take the velocity of electromagnetic radiation to be 2.998×10^8 m s^{-1}.

Chemical background

This type of discharge is typically seen in certain forms of street lighting.

When excited by an electric discharge, sodium vapour produces an emission spectrum which includes a yellow line at a wavelength of 589 nm. On closer analysis, this is seen to consist of two very closely spaced lines (a doublet) of wavelengths 589.76 nm and 589.16 nm respectively.

Solution to worked example

We need to calculate the value of the quantity $1/\lambda$. If we start with the defining equation

$$c = \lambda v$$

we can divide both sides by v to give

$$\frac{c}{v} = \lambda$$

As we wish to calculate $1/\lambda$, we now need to take the reciprocal of each side of this equation. The reciprocal of λ is $1/\lambda$ while the reciprocal of any fraction is obtained by 'turning it upside down'. This now gives us

$$\frac{1}{\lambda} = \frac{v}{c}$$

$$= \frac{5.086 \times 10^8 \text{ MHz}}{2.998 \times 10^8 \text{ m s}^{-1}}$$

The first thing to notice is that both the top and the bottom of this fraction contain a 10^8 term, which can be cancelled. We then need to rewrite the unit MHz in a form which is of more use to us. The basic frequency unit Hz is equal to s^{-1}, while the prefix M means 10^6. This now gives

$$\frac{1}{\lambda} = \frac{5.086 \times 10^6 \, s^{-1}}{2.998 \, m \, s^{-1}}$$

The s^{-1} unit now appears in both the top and bottom of the right side of this equation, and can be cancelled. Using a calculator we now obtain

$$\frac{1}{\lambda} \approx 1.696 \times 10^6 \, m^{-1} \quad \boxed{\boxplus}$$

While this is an acceptable unit for wavenumber, it is more usual to use units of cm^{-1}. We can do this if we realize that

$$100 \, cm = 1 \, m$$

or more usefully here

$$10^2 \, cm = 1 \, m$$

We simply replace m in our expression with 10^2 cm. This gives

$$\frac{1}{\lambda} = 1.696 \times 10^6 \, m^{-1}$$

$$= 1.696 \times 10^6 \, (10^2 \, cm)^{-1}$$

The power of -1 is applied to every term in the bracket, so we need to use the fact that $(10^2)^{-1} = 10^{-2}$ to give

$$\frac{1}{\lambda} = 1.696 \times 10^6 \times 10^{-2} \, cm^{-1}$$

$$= 1.696 \times 10^4 \, cm^{-1}$$

7.2 The Beer–Lambert Law

When radiation is incident upon some medium, it will be absorbed to a certain extent, as shown in Figure 7.3. This will result in its intensity being reduced, and

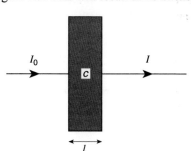

Figure 7.3   Absorption of radiation.

The distance l is commonly known as the path length.

such behaviour is described by the Beer–Lambert Law which can be expressed by the equation

$$A = \varepsilon c l$$

where ε is a constant known as the absorption coefficient, c is the concentration of the absorbing medium and l is the distance through which the light travels within the medium. The quantity A is known as the absorbance, and is defined as

$$A = \log \frac{I_0}{I}$$

where I_0 is the intensity of the incident radiation and I is the intensity once it has passed through the medium.

7.2.1   Rules of logarithms

These were discussed in section 4.4.2.

Worked example 7.3

The fraction of incident radiation transmitted is the transmittance T which is defined by the equation

$$T = \frac{I}{I_0}$$

Obtain an equation which relates the percentage transmittance T_{100} to the absorbance A.

Chemical background

Spectrophotometers are generally designed to work in one region of the spectrum only. Specific wavelengths are produced by using a rotating prism or a diffraction grating.

Many spectrophotometers are capable of displaying both absorbance and transmittance, and the relationship between these quantities is not immediately obvious. Generally, absorbance is the preferred quantity because of its proportionality to both concentration and path length.

Solution to worked example

The percentage transmittance T_{100} is related to the transmittance T (a fractional value) by the simple expression

$$T_{100} = 100T$$

$$= 100 \frac{I}{I_0}$$

This rearranges to give

$$\frac{I_0}{I} = \frac{100}{T_{100}}$$

which can be substituted into the expression for A

$$A = \log \frac{I_0}{I} = \log \frac{100}{T_{100}}$$

Since we know that

$$\log\left(\frac{X}{Y}\right) = \log X - \log Y$$

we can rearrange the right-hand side of this equation to give

$$A = \log 100 - \log T_{100}$$

and since $\log 100 = 2$ we obtain

$$A = 2 - \log T_{100}$$

which finally can be rearranged to give

$$\log T_{100} = 2 - A$$

7.3 Rotational spectroscopy

Rotating molecules give rise to spectra in the microwave region. It is possible to define three mutually perpendicular principal axes of rotation which pass through the centre of gravity of a molecule, and there are consequently three corresponding moments of inertia which depend on the molecular shape. The treatment of rotational spectra is consequently different for linear molecules (such as HCl), symmetric tops (such as CH_3F), spherical tops (such as CH_4) and asymmetric tops (such as H_2O). These different types of molecule are shown in Figure 7.4.

7.3.1 Sequences

This subject was discussed in section 6.3.2.

Worked example 7.4

The rotational energy levels $F(J)$ of a rigid diatomic molecule are given by the expression

$$F(J) = BJ(J + 1)$$

where J is the rotational quantum number (taking values 0, 1, 2, ...) and B is the rotational constant. What are the three lowest rotational energy levels, in terms of B?

Each energy level has a degeneracy of $(2J + 1)$; in other words, there are $(2J + 1)$ levels having each energy value. The population of each level of energy $F(J)$ is proportional to $(2J + 1) \exp(-F(J)/kT)$.

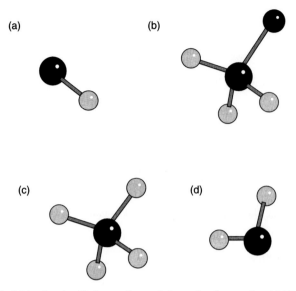

Figure 7.4   Molecules classified according to their rotational properties: (a) HCl (linear); (b) CH_3F (symmetric top); (c) CH_4 (spherical top); and (d) H_2O (asymmetric top).

Chemical background

This problem shows that the rotational energy levels in a diatomic molecule are not evenly spaced. However, if we consider the differences between levels then we do see an even spacing, and this is apparent in rotational spectra.

Solution to worked example

To obtain the three lowest energy levels, we need to substitute the three lowest values of J into the given expression. We then have

$$F(0) = (B \times 0)(0 + 1) = 0$$

$$F(1) = (B \times 1)(1 + 1) = 2B$$

$$F(2) = (B \times 2)(2 + 1) = 6B$$

as shown in Figure 7.5.

Worked example 7.5

The rotational energy levels $F(J)$ of a rigid diatomic molecule are given by the expression

$$F(J) = BJ(J + 1)$$

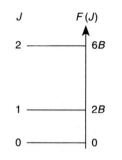

Figure 7.5 Spacing of rotational energy levels.

where J is the rotational quantum number and B is the rotational constant. Obtain an expression for the energy of a rotational transition, given that such transitions are subject to the selection rule $\Delta J = \pm 1$.

Chemical background

The rotational constant B can be calculated from the equation

$$B = \frac{h^2}{8\pi^2 I}$$

where h is Planck's constant and I is the moment of inertia. The selection rule is obtained from the application of the Schrödinger equation.

A rotational spectrum will only be seen in the case of heteronuclear diatomic molecules. These will be able to interact with the radiation as their dipole moment changes during the rotation. No such spectrum would be seen for Cl_2 but it would for HCl.

Solution to worked example

To calculate the energy of a transition, we need to obtain expressions for the energies of two adjacent levels. We choose these to have rotational quantum numbers J and $J + 1$ respectively. We then have

$$F(J) = BJ(J + 1)$$

as given above.

To obtain $F(J + 1)$, we need to replace each occurrence of J in the above equation by $J + 1$. This gives

$$F(J + 1) = B(J + 1)((J + 1) + 1)$$

$$= B(J + 1)(J + 2)$$

We multiply the two brackets together by considering the four combinations of terms to obtain

$$F(J + 1) = B(J^2 + 2J + J + 2)$$

$$= B(J^2 + 3J + 2)$$

The energy difference $\Delta F(J)$ is given by

$$\Delta F(J) = F(J + 1) - F(J)$$

$$= B(J^2 + 3J + 2) - BJ(J + 1)$$

$$= B(J^2 + 3J + 2 - J^2 - J)$$

$$= B(2J + 2)$$

We can remove the common factor of 2 from within the bracket to leave

$$\Delta F(J) = 2B(J + 1)$$

7.4
Vibrational
spectroscopy

The starting point for a study of vibrational spectroscopy is usually the model of a simple harmonic oscillator, a system which is often treated in some detail in courses on quantum mechanics. Any energy possessed by this system over and above its equilibrium value is due to extension or compression of the bond. The vibrational frequency, ω_e, of this system is given by the formula

$$\omega_e = \frac{1}{2\pi} \sqrt{\frac{k}{\mu}}$$

where k is the force constant and μ is a quantity known as the reduced mass of the system, which is related to the masses of the atoms forming the bonds.

Worked example 7.6

The vibrational energy $G(v)$ of the harmonic oscillator is given by the equation

$$G(v) = (v + \tfrac{1}{2})\omega_e$$

where v is the vibrational quantum number (having values 0, 1, 2, ...). Calculate the energy of the vibrational level having $v = 2$ for the HCl molecule. Its force constant is 516 N m^{-1}, and its reduced mass is given by the equation

$$\mu = \frac{m_H m_{Cl}}{m_H + m_{Cl}}$$

where m represents the mass of the specified atom.

One way of remembering the correct formula for calculating the reduced mass is to consider the units. In the formula here, we have units of mass^2 divided by units of mass; the reciprocal would clearly give incorrect units.

Chemical background

In practice, there will be very few molecules in this vibrational level, so it will not contribute significantly to the vibrational spectrum.

Solution to worked example

For 1 mol of HCl we would have

$$\mu = \frac{35.453 \text{ g mol}^{-1} \times 1.0079 \text{ g mol}^{-1}}{35.453 \text{ g mol}^{-1} + 1.0079 \text{ g mol}^{-1}}$$

$$\simeq \frac{35.733 \text{ g mol}^{-1}}{36.461}$$

$$\simeq 0.98003 \text{ g mol}^{-1}$$

and for one molecule we need to divide by Avogadro's constant:

$$\mu = \frac{0.9800 \text{ g mol}^{-1}}{6.022 \times 10^{23} \text{ mol}^{-1}}$$

$$\simeq \frac{0.1627 \text{ g mol}^{-1}}{10^{23} \text{ mol}^{-1}}$$

$$= 1.627 \times 10^{-24} \text{ g}$$

$$= 1.627 \times 10^{-27} \text{ kg}$$

so from the formula above the fundamental frequency can be calculated as

$$\omega_e = \frac{1}{2\pi} \sqrt{\frac{k}{\mu}}$$

$$= \frac{1}{2\pi} \sqrt{\frac{516 \text{ N m}^{-1}}{1.627 \times 10^{-27} \text{ kg}}}$$

Since $1 \text{ N} = 1 \text{ kg m s}^{-2}$ this becomes

$$\omega_e = \frac{1}{2\pi} \sqrt{\frac{516 \text{ kg m s}^{-2} \text{ m}^{-1}}{1.627 \times 10^{-27} \text{ kg}}}$$

Cancelling units (kg top and bottom, and m and m^{-1}) then gives

$$\omega_e = \frac{1}{2\pi} \sqrt{\frac{516 \text{ s}^{-2}}{1.627 \times 10^{-27}}}$$

$$\simeq \frac{\sqrt{3.171 \times 10^{29} \text{ s}^{-2}}}{2\pi}$$

$$\simeq \frac{5.631 \times 10^{14} \text{ s}^{-1}}{2\pi}$$

$$\simeq 8.962 \times 10^{13} \text{ s}^{-1}$$

This answer is easily transformed into other quantities by the use of appropriate units. Multiplication by Planck's constant h will give energy (in J) and division by the velocity of light c will give wavenumber (in m^{-1}).

We now need to substitute the value of $v = 2$ into the expression for the vibrational energy. This gives us

$$G(2) = (2 + \tfrac{1}{2})\omega_e$$

$$= 2.5 \times 8.962 \times 10^{13}\ s^{-1}$$

$$\approx 2.241 \times 10^{14}\ s^{-1}$$

$$= 2.241 \times 10^{14}\ Hz$$

since $1\ Hz = 1\ s^{-1}$.

Worked example 7.7

The vibrational energy $G(v)$ of the harmonic oscillator is given by the equation

$$G(v) = (v + \tfrac{1}{2})\omega_e$$

where v is the vibrational quantum number. Given that the selection rule for a vibrational transition is $\Delta v = \pm 1$, obtain an expression for the energy associated with such a transition.

Chemical background

The harmonic oscillator approximation is valid for vibrations which displace the bond length by up to about 10% of its equilibrium value. Above this, it is better to use the model of an anharmonic oscillator, as used in Problem 3 at the end of this chapter.

The energy associated with this transition turns out to be a constant value, equal to the vibrational frequency of the system. This is because the vibrational energy levels are equally spaced and the vibrating molecule will only absorb energy of the same frequency as its own natural vibrations.

Solution to worked example

As in the rotational case in Worked Example 7.4, we need to set up expressions for the energy of two adjacent levels. We have from the question

$$G(v) = (v + \tfrac{1}{2})\omega_e$$

Replacing every occurrence of v by $v + 1$ gives

$$G(v + 1) = ((v + 1) + \tfrac{1}{2})\omega_e$$

$$= (v + 1 + \tfrac{1}{2})\omega_e$$

$$= (v + \tfrac{3}{2})\omega_e$$

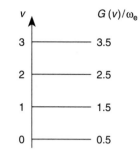

Figure 7.6   Spacing of vibrational energy levels.

The energy $\Delta G(v)$ associated with this vibrational transition will now be given by

$$\Delta G(v) = G(v + 1) - G(v)$$

$$= (v + \tfrac{3}{2})\omega_e - (v + \tfrac{1}{2})\omega_e$$

$$= (v + \tfrac{3}{2} - v - \tfrac{1}{2})\omega_e$$

$$= \omega_e$$

The spacing of vibrational energy levels is shown in Figure 7.6.

7.5 Rotation–vibration spectroscopy

It is reasonable to suppose that a molecule will be undergoing both rotational and vibrational motion at a given time, so it is not surprising that it is possible to investigate the coupling of these motions by means of spectroscopic techniques. The Born–Oppenheimer approximation states that it is possible to separate the vibrational and rotational motion, and it is therefore possible to simply add the contributions from each to obtain the vibration–rotation energy. Figure 7.7 shows the rotational levels superimposed on the vibrational levels for a molecule.

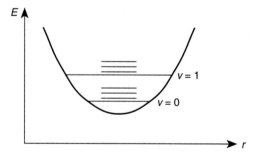

Figure 7.7   Vibrational and rotational energy levels on the same scale.

Worked example 7.8

The energy $E(v, J)$ due to combined vibration and rotation is given by the equation

$$E(v, J) = BJ(J + 1) + (v + \tfrac{1}{2})\omega_e$$

where v and J are the vibrational and rotational quantum numbers respectively. Given the selection rules $\Delta v = \pm 1$ and $\Delta J = \pm 1$, obtain an expression for the energy due to simultaneous vibrational and rotational transitions.

Chemical background

It is important to realize that the separation of the rotational levels is much less than that of the vibrational levels. Typical values are 0.05 kJ mol^{-1} for rotational levels and 10 kJ mol^{-1} for vibrational levels. These compare with a separation of the order of 500 kJ mol^{-1} for electronic levels.

The selection rules given show that it is only possible for a vibrational change to occur if a rotational change takes place simultaneously. It is possible to have $\Delta v = 0$, which represents the pure rotational change we met in Worked Example 7.4, but it is not possible to have $\Delta J = 0$ representing a vibrational change alone.

Solution to worked example

We proceed as in the previous problems by considering the difference between two energy levels, remembering this time that we need to deal with two simultaneous changes. The notation usually employed is that representing a transition from a lower vibrational level v'' and rotational level J'' to an upper vibrational level v' and corresponding rotational level J'. We then have

$$\Delta E(v, J) = E(v', J') - E(v'', J'')$$

$$= BJ'(J' + 1) + (v' + \tfrac{1}{2})\omega_e - BJ''(J'' + 1) - (v'' + \tfrac{1}{2})\omega_e$$

$$= BJ'(J' + 1) - BJ''(J'' + 1) + (v' - v'')\omega_e$$

Since we have the selection rule $\Delta v = \pm 1$ it follows that

$$v' - v'' = 1$$

since v' is the higher level and so

$$\Delta E(v, J) = BJ'(J' + 1) - BJ''(J'' + 1) + \omega_e$$

When we move to the higher vibrational level, it is possible for the change in rotational quantum number to be either positive or negative. If

$$J' = J'' + 1$$

we have

$$\Delta E(J') = BJ'(J' + 1) - B(J' - 1)J' + \omega_e$$
$$= BJ'^2 + BJ' - BJ'^2 + BJ' + \omega_e$$
$$= \omega_e + 2BJ'$$

Alternatively, if

$$J' = J'' - 1$$

then

$$\Delta E(J'') = B(J'' - 1)J'' - BJ''(J'' + 1) + \omega_e$$
$$= BJ''^2 - BJ'' - BJ''^2 - BJ'' + \omega_e$$
$$= \omega_e - 2BJ''$$

Note that either of these expressions could be obtained using J' or J''.

7.6
Nuclear magnetic resonance spectroscopy

This form of spectroscopy is one of the most powerful analytical tools used by organic chemists. While spectra for complicated molecules can be quite difficult to unravel, the underlying mathematics are quite simple to understand. Such spectra are basically due to the splitting of the proton spin energy levels by a magnetic field. Peaks are classified by their chemical shifts (measured in parts per million, ppm) relative to the standard tetramethyl silane, $SiMe_4$.

Chemical shifts can be measured on the τ scale, on which tetramethylsilane has a value of 10. The chemical shifts of most groups of interest then fall between 0 and 10, with a phenyl hydrogen having a value of 2.7, for example.

7.6.1   Pascal's Triangle

If we wish to expand the expression $(x + 2)^2$ we could rewrite it as

$$(x + 2)^2 = (x + 2)(x + 2)$$

and multiply out the brackets to give

$$(x + 2)^2 = x(x + 2) + 2(x + 2)$$
$$= x^2 + 2x + 2x + 4$$
$$= x^2 + 4x + 4$$

Similarly, $(x + 2)^3$ could be calculated as

$$(x + 2)^3 = (x + 2)(x + 2)^2$$
$$= (x + 2)(x^2 + 4x + 4)$$
$$= x(x^2 + 4x + 4) + 2(x^2 + 4x + 4)$$
$$= x^3 + 4x^2 + 4x + 2x^2 + 8x + 8$$
$$= x^3 + 6x^2 + 12x + 8$$

In a similar fashion we could obtain expressions for $(x + 2)^4$ and higher powers, but this is obviously going to become more and more difficult as the power increases.

Fortunately, such step-by-step expansion is not necessary if we use a device known as **Pascal's Triangle**. This is shown in Figure 7.8. You will be relieved to know that this does not need to be memorized, as it can be set up using a few simple rules which are easily remembered.

We start off on the top line with the number 1.

$$1$$

This will actually be of more use to us if we write it with zeros on either side as

$$0 \quad 1 \quad 0$$

To obtain the second line, we now add each pair of numbers in the line above, and write the result in the speace underneath them. This gives

$$0 \underset{+}{\smile} 1 \underset{+}{\smile} 0$$
$$1 \quad 1$$

We now add the zeros on either side of this second line:

$$0 \quad 1 \quad 0$$

$$0 \quad 1 \quad 1 \quad 0$$

$$1$$

$$1 \qquad\qquad 1$$

$$1 \qquad 2 \qquad 1$$

$$1 \qquad 3 \qquad 3 \qquad 1$$

$$1 \qquad 4 \qquad 6 \qquad 4 \qquad 1$$

$$1 \qquad 5 \qquad 10 \qquad 10 \qquad 5 \qquad 1$$

$$1 \qquad 6 \qquad 15 \qquad 20 \qquad 15 \qquad 6 \qquad 1$$

Figure 7.8   Pascal's Triangle.

To obtain the third line, we repeat the process of adding each pair of numbers in the second line and writing the result in the space underneath. We then have

$$0 \quad 1 \quad 0$$

$$0 \underset{+}{} 1 \underset{+}{} 1 \underset{+}{} 0$$

$$0 \quad 1 \quad 2 \quad 1 \quad 0$$

The rest of Pascal's Triangle can be built up in the same way.

How does this help us to expand brackets to a specified power? From the examples given, we can see that when a bracket such as $(x + 2)$ is raised to some power, we obtain a polynomial expression. These were discussed in section 2.2.2, where we saw that each term in x is multiplied by a constant called the coefficient. For example, in a term such as $3x^4$ we say that the coefficient of x^4 is 3. It is these coefficients that can be determined by using Pascal's Triangle.

From Figure 7.8, we see that the first line of Pascal's Triangle gives us the coefficient of a zero-order expansion (power 0), the next line a first-order expansion (power 1) and so on. Therefore to evaluate $(x + 2)^2$ we need the coefficients from the third line, which are 1, 2, 1. As we saw earlier, we would expect this expansion to give us terms in x^2 and x and a constant term, so we have

$$\text{coefficient of } x^2 = 1$$

$$\text{coefficient of } x = 2$$

$$\text{coefficient of constant} = 1$$

In fact, this expansion can be expressed as

$$(x + 2)^2 = 1 \times x^2 \times 2^0 + 2 \times x^1 \times 2^1 + 1 \times x^0 \times 2^2$$

where each term comprises the appropriate coefficient from Pascal's Triangle, a term in x and a term in 2. Notice that as the power of x decreases that of 2 increases, but the two powers added together are always equal to 2. If we remember that any number raised to a zero power is 1, and any number raised to power 1 is itself, this expression simplifies to give

$$(x + 2)^2 = 1 \times x^2 \times 1 + 2 \times x \times 2 + 1 \times 1 \times 4$$

$$= x^2 + 4x + 4$$

which is, of course, the result we obtained before.

Let us now consider an example which would be much more difficult to evaluate without Pascal's Triangle, $(2x + 3)^4$. From Figure 7.8 we see that the appropriate coefficients for an expansion of degree 4 are

$$1 \quad 4 \quad 6 \quad 4 \quad 1$$

This time we need to consider terms in $2x$ and 3, so we have

$$(2x + 3)^4 = 1 \times (2x)^4 \times 3^0 + 4 \times (2x)^3 \times 3^1$$
$$+ 6 \times (2x)^2 \times 3^2 + 4 \times (2x)^1 \times 3^3 + 1 \times (2x)^0 \times 3^4$$

Notice that when we raise a bracket to the fourth power we have five terms. The power of $(2x)$ decreases while that of 3 increases and the sum of these powers is always 4. It is important to realize that, for example,

$$(2x)^4 = 2^4 x^4 = 16x^4$$

so the power is applied to *both* quantities within the bracket. The expansion can then be simplified to

$$(2x + 3)^4 = 1 \times 16x^4 \times 1 + 4 \times 8x^3 \times 3 + 6 \times 4x^2 \times 9$$
$$+ 4 \times 2x \times 9 + 1 \times 1 \times 81$$
$$= 16x^4 + 96x^3 + 216x^2 + 72x + 81$$

Worked example 7.9

The peaks due to each group of protons in a molecule are split into $n + 1$ by an adjacent group of n protons, the intensity of these peaks being given by the coefficients of Pascal's Triangle. What is the distribution of peaks within the CH_2 and CH_3 signals of ethanol, ignoring any interaction with the OH group?

Chemical background

The reason for ignoring any interaction with the OH group is that its proton undergoes rapid exchange with other molecules, due to the presence of small amounts of acid, and so it is unable to interact with protons in the other groups of the molecule.

Solution to worked example

Nuclear magnetic resonance spectra are most commonly measured for the ^1H nucleus. Other nuclei which can be used include ^{19}F, ^{31}P and, since the advent of Fourier transform methods, ^{13}C.

The signal due to the CH_3 group will be split into three $(2 + 1)$ peaks by the CH_2 group, and from Pascal's Triangle these will be in the ratio $1:2:1$. The signal due to the CH_2 group will be split into four $(3 + 1)$ peaks by the CH_3 group, in the ratio $1:3:3:1$. These peaks are shown in Figure 7.9.

Exercises

1. Identify the type of proportion between the appropriate functions of x and y in the relationships:

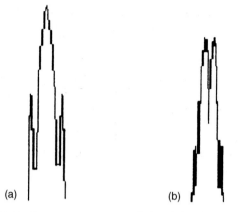

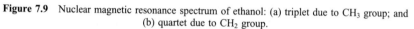

Figure 7.9   Nuclear magnetic resonance spectrum of ethanol: (a) triplet due to CH_3 group; and (b) quartet due to CH_2 group.

(a) $x \ln y = K$

(b) $\dfrac{x^2}{y} = K$

where K is a constant.

2. Identify the proportionality constant when y is written as a function of x for the relationships:

   (a) $ax = by$

   (b) $\left(\dfrac{x}{a}\right)^2 = \left(\dfrac{b}{y}\right)$

3. Write each of these expressions as a single logarithm.

   (a) $\ln x^3 - \ln x^2$

   (b) $\ln xy + \ln x^2$

4. Write each of these expressions as a sum or difference of logarithms.

   (a) $\ln \sqrt{\dfrac{x+1}{x-1}}$

   (b) $\ln 5x^4$

5. The terms in a series are defined by the expression

$$f(n) = 3n^3 + 4n^2 - 2n + 1$$

What is the value of $f(2) - f(1)$?

6. The terms in a series are defined by the expression

$$g(N) = N^2 + 2N$$

Obtain an expression for $g(N+1) - g(N)$.

7. Verify that the sequence

$$1, \frac{3}{2}, \frac{7}{4}, \frac{15}{8}, \frac{31}{16}, \ldots$$

can be represented as

$$f(n) = 2 - \frac{1}{2^{n-1}}$$

What is the limit of this sequence as n tends to infinity?

8. Use Pascal's Triangle to obtain expressions for

   (a) $(x + 3)^4$

   (b) $(2x - 4)^5$

9. In the expansion of $(x - 3y)^4$, what is the coefficient of y^3?

10. Use Pascal's Triangle to obtain an expression for $(1 - x)^6$. By setting x to 0.002, obtain the value of 0.998^6 to 6 significant figures *without* using a calculator.

Problems

1. Lambert's Law states that each successive layer of thickness dx of a medium absorbs an equal fraction

$$-\frac{dI}{I}$$

of radiation of intensity I, so that

$$-\frac{dI}{I} = b \, dx$$

where b is a constant. Obtain an expression for the intensity I of light which has travelled through a distance l, assuming that $I = I_0$ when $l = 0$.

2. The rotational energy levels $F(J)$ of a non-rigid molecule are given by the formula

$$F(J) = BJ(J + 1) - DJ^2(J + 1)^2$$

where J is the rotational quantum number, B the rotational constant and D is the centrifugal distortion constant. Obtain an expression for

$$F(J + 1) - F(J)$$

in terms of B and D.

3. The vibrational energy levels $G(v)$ of an anharmonic oscillator are given by the expression

$$G(v) = (v + \tfrac{1}{2})\omega_e - (v + \tfrac{1}{2})^2 \omega_e x_e$$

where v is the vibrational quantum number, ω_e is the vibrational frequency and x_e is the anharmonicity constant. Obtain an expression for $G(1) - G(0)$, which is the most likely vibrational transition since higher levels will only be very sparsely populated.

4. The wavenumbers $\tilde{v}$ of lines in the rotation–vibration spectrum of carbon monoxide form a series defined by

$$\frac{\tilde{v}}{cm^{-1}} = 2143.28 + 3.813m - 0.0175m^2$$

where $m = \ldots, -2, -1, 0, 1, 2, \ldots$.

Obtain an expression for $\tilde{v}(m + 1) - \tilde{v}(m)$ and show that for positive values of m the line spacing decreases while for negative values it increases with increasing magnitude of m.

5. The peaks due to each group of protons in a molecule are split into $n + 1$ by an adjacent group of n protons, the intensities of these peaks being given by the coefficients of Pascal's Triangle. Describe the appearance of the nuclear magnetic resonance spectrum of isopropyl iodide, $(CH_3)_2CHI$.

8 Statistical mechanics

Thermodynamics only deals with a macroscopic picture of matter, and its principles have all been derived by observing experimental behaviour. In contrast, quantum mechanics is concerned with the microscopic view and it starts from a more theoretical point, even though many of its results do explain experimental behaviour. Invariably, in quantum mechanics we are considering only single atoms or molecules.

The topic of statistical mechanics provides a link between these subjects. We start with the microscopic view of matter and apply statistical techniques to a large number of chemical entities in order to reproduce the macroscopic functions of thermodynamics. This requires a few mathematical techniques which we have not yet come across, as well as some of those met in earlier chapters.

8.1
Kinetic energy

Chemical particles can possess two types of energy: potential and kinetic. Potential energy is due to their interactions with other particles through intermolecular and intramolecular forces, while kinetic energy is due to the fact that these particles have motion. We saw in Chapter 7 how to express the rotational and vibrational energies mathematically, but molecules also have translational energy of motion. It is this motion which causes a molecule to possess kinetic energy.

8.1.1   Straight line graphs

This topic was covered in section 2.4.2.

Worked example 8.1

The kinetic energy ε of a molecule having mass m and velocity v is given by the equation

$$\varepsilon = \tfrac{1}{2}mv^2$$

(a) Identify the variables and constants in this expression.
(b) What information could be obtained from a set of measurements of the variables in the expression?

Chemical background

It is possible to identify the components of the kinetic energy and the velocity in each of the three directions x, y and z for a particle. We can therefore have

$$\varepsilon = \varepsilon_x + \varepsilon_y + \varepsilon_z$$

where ε is the total kinetic energy of the particle and ε_x, ε_y and ε_z are its components in the three directions. The fact that the energy can be apportioned in this way is an example of the Principle of Equipartition of Energy.

The Principle of Equipartition of Energy fails for vibrational energy at lower temperatures.

Solution to worked example

(a) The mass m of a given particle will be constant so that the variable ε is a function of the variable v.
(b) If we have a set of measurements of ε and v, we will obtain most information if we are able to plot a straight line graph. If we compare the expression

$$\varepsilon = \tfrac{1}{2}mv^2$$

with the general equation of a straight line

$$y = mx + c$$

we can see that if we plot v^2 on the x-axis and ε on the y-axis we will obtain a straight line with gradient $\tfrac{1}{2}m$ which passes through the origin (since c is zero).

The statement in part (a) is true as long as we are able to ignore relativistic effects. Otherwise, the mass will be a function of velocity. In practice, this is only a problem when the velocity is very high.

8.2 Configurations

Since statisticstical mechanics involves the application of statistics to an arrangement of chemical particles, we need to be able to describe the way in which such particles are arranged (their configuration). The number of complexions is the number of ways in which a specified number of particles can be arranged in each energy level. For example, we might wish to specify two molecules in the lowest energy level, three in the next level and four in the level above. Such an arrangement can only be achieved in a certain number of ways starting with nine molecules.

Strictly speaking, we are dealing with instantaneous configurations, which will be constantly changing.

8.2.1 Factorials

A factorial is a number written with an exclamation mark (!) immediately after it, such as 2!, 45! or 8! This strange looking notation has a simple meaning, as illustrated by the following examples

$$5! = 5 \times 4 \times 3 \times 2 \times 1 = 120$$

$$9! = 9 \times 8 \times 7 \times 6 \times 5 \times 4 \times 3 \times 2 \times 1 = 362\,880$$

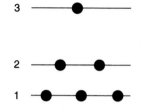

Figure 8.1   Distribution of molecules in the system described in Worked Example 8.2.

In general, for a number n we have

$$n! = n \times (n-1) \times (n-2) \times (n-3) \cdots 3 \times 2 \times 1$$

Worked example 8.2

The number of complexions Ω is given by the equation

$$\Omega = \frac{N!}{\prod_i n_i!}$$

where N is the total number of particles and n_i is the number of particles in energy level i. The symbol $\prod$ denotes that we calculate the product of the quantities specified. What is the number of complexions if 3 molecules are in level 1, 2 molecules in level 2 and 1 molecule in level 3, as shown in Figure 8.1?

Chemical background

The arrangement of molecules between these energy levels is known as a state, and will be constantly changing. Normally, the broadest distribution of the molecules across the energy levels will predominate.

Solution to worked example

The first stage is to write out the quantities which we have been given in the question using the notation of the given equation. If there are n_i molecules in level i, we can write

$$n_1 = 3 \quad n_2 = 2 \quad n_3 = 1$$

and the total number N of molecules is then given as

$$N = n_1 + n_2 + n_3$$
$$= 3 + 2 + 1$$
$$= 6$$

We can now calculate $N!$.

$$N! = 6!$$

$$= 6 \times 5 \times 4 \times 3 \times 2 \times 1$$

$$= 720$$

The other factorials required are

$$n_1! = 3! = 3 \times 2 \times 1 = 6$$

$$n_2! = 2! = 2 \times 1 = 2$$

and

$$n_3! = 1! = 1$$

We can now substitute into our equation to give

$$\Omega = \frac{N!}{\prod_i n_i!} = \frac{N!}{n_1! \, n_2! \, n_3!}$$

$$= \frac{6!}{3! \, 2! \, 1!} = \frac{720}{6 \times 2 \times 1}$$

$$= \frac{720}{12}$$

$$= 60$$

There is, however, a slightly neater way of performing this calculation. It will be shown here since the technique can be of use when we are dealing with symbols rather than actual numbers.

Since

$$6! = 6 \times 5 \times 4 \times 3 \times 2 \times 1$$

and

$$3! = 3 \times 2 \times 1$$

we can write

$$6! = 6 \times 5 \times 4 \times 3!$$

If we now substitute into the equation for Ω, we obtain

$$\Omega = \frac{6!}{3! \, 2! \, 1!}$$

$$= \frac{6 \times 5 \times 4 \times 3!}{3! \, 2! \, 1!}$$

Cancelling 3! top and bottom, and writing 2! as 2, and 1! as 1 gives

$$\Omega = \frac{6 \times 5 \times 4}{2}$$

$$= 6 \times 5 \times 2$$

$$= 60$$

This second method makes the calculation easier when a calculator is not available.

8.3
The Boltzmann equation

It is not always appreciated that the Boltzmann distribution is given in terms of general energy levels, while the Maxwell–Boltzmann distribution relates only to kinetic energy and molecular speeds.

The Boltzmann distribution law allows us to specify the distribution of molecules among the energy levels. An increase in temperature results in an increase in molecular energy. Therefore, we would expect such a function to depend upon temperature.

The essential process in its determination is to maximize the number of complexions Ω with respect to the number of particles in each level n_i. If Ω is a maximum, it follows that $\ln \Omega$ will also be a maximum, so that the derivative which needs to be calculated and set equal to zero is

$$\frac{\partial \ln \Omega}{\partial n_i}$$

8.3.1 Differentiation of logarithms

This topic was covered in section 4.5.2.

8.3.2 Differentiation of products

This subject was described in section 6.5.1.

Worked example 8.3

Stirling's approximation states that

$$\ln n! = n \ln n - n$$

for large values of n. In the derivation of the Boltzmann equation, this leads to the equation

$$\frac{\partial \ln \Omega}{\partial n_i} = -\frac{\partial}{\partial n_i}(n_i \ln n_i - n_i)$$

Obtain an expression for the derivative on the right-hand side of this equation.

Chemical background

Stirling's approximation is very useful when we are dealing with large numbers, as is the case in statistical mechanics. Consider how difficult it would be to calculate the factorial of Avogadro's number. If you try it on your calculator you will see that it cannot be done, but it is straightforward using the equation above.

Solution to worked example

The expression to be differentiated consists of two terms which will be considered in turn. The first involves determining

$$\frac{\partial}{\partial n_i}(n_i \ln n_i)$$

Remember that the rule for differentiating a product is to differentiate the first function and multiply by the second, then differentiate the second function and multiply by the first. This will give us

$$\frac{\partial}{\partial n_i}(n_i \ln n_i) = (\ln n_i)\frac{\partial}{\partial n_i}(n_i) + n_i\frac{\partial}{\partial n_i}(\ln n_i)$$

Since

$$\frac{\partial}{\partial n_i}(n_i) = 1 \quad \text{and} \quad \frac{\partial}{\partial n_i}(\ln n_i) = \frac{1}{n_i}$$

we obtain

$$\frac{\partial}{\partial n_i}(n_i \ln n_i) = \ln n_i + n_i\frac{1}{n_i}$$

$$= \ln n_i + 1$$

and so

$$\frac{\partial}{\partial n_i}\left(n_i \ln n_i - n_i\frac{\partial \ln \Omega}{\partial n_i}\right) = -(\ln n_i + 1 - 1)$$

$$= -\ln n_i$$

Worked example 8.4

The Boltzmann equation for the number of particles n_i in an energy level i having energy ε_i is

$$\frac{n_i}{N} = \frac{\exp\left(\dfrac{-\varepsilon_i}{kT}\right)}{\sum\limits_i \exp\left(\dfrac{-\varepsilon_i}{kT}\right)}$$

where N is the total number of particles and k is the Boltzmann constant.

Obtain an expression for the number of particles in level i relative to that in level j.

Chemical background

The quantity

$$\sum_i \exp\left(\frac{-\varepsilon_i}{kT}\right)$$

in this expression is known as the partition function and is usually given the symbol q. This is of great importance in statistical mechanics. We often calculate the partition functions for translation, vibration, rotation and electrons separately.

Solution to worked example

We can use the equation as given

$$\frac{n_i}{N} = \frac{\exp\left(\dfrac{-\varepsilon_i}{kT}\right)}{\sum_i \exp\left(\dfrac{-\varepsilon_i}{kT}\right)}$$

for level i, and rewrite it to give

$$\frac{n_j}{N} = \frac{\exp\left(\dfrac{-\varepsilon_j}{kT}\right)}{\sum_i \exp\left(\dfrac{-\varepsilon_i}{kT}\right)}$$

for level j. Notice that we are still considering energy levels i in the second expression; it really does not matter what symbol we use as long as that on the summation sign $\sum$ corresponds to that inside the summation (in this case on the energy level).

When we take the ratio of n_i and n_j we obtain

$$\frac{n_i}{n_j} = \frac{\exp\left(\dfrac{-\varepsilon_i}{kT}\right)}{\exp\left(\dfrac{-\varepsilon_j}{kT}\right)}$$

since the denominator of the defining equations is the same for n_i and n_j. To simplify this expression, we need to use the fact that

$$\frac{a^x}{a^y} = a^{x-y}$$

so we obtain

$$\frac{n_i}{n_j} = \exp\left(\frac{-\varepsilon_i}{kT} - \frac{-\varepsilon_j}{kT}\right)$$

We can rewrite

$$-\varepsilon_i - (-\varepsilon_j)$$

as

$$-\varepsilon_i + \varepsilon_j$$

which is given more usually as

$$-(\varepsilon_i - \varepsilon_j)$$

and so the expression we obtain is

$$\frac{n_i}{n_j} = \exp\left(\frac{-(\varepsilon_i - \varepsilon_j)}{kT}\right)$$

8.4
The partition function

In the previous section, we met the idea of a partition function q, and saw that it was defined as

$$q = \sum_i \exp\left(\frac{-\varepsilon_i}{kT}\right)$$

The partition function q defined here is actually the molecular partition function. The partition function Q for the whole system is equal to q^N, and this needs to be divided by $N!$ if the particles making up the system are indistinguishable, such as in a gas.

8.4.1   Discontinuities

Discontinuous functions were discussed in section 3.3.1.

Worked example 8.5

The vibrational partition function q_v is defined by the equation

$$q_v = \frac{1}{1 - \exp(-hv/kT)}$$

where h is Planck's constant, v is the vibrational frequency, k is Boltzmann's constant and T is the absolute temperature. When will this function be discontinuous?

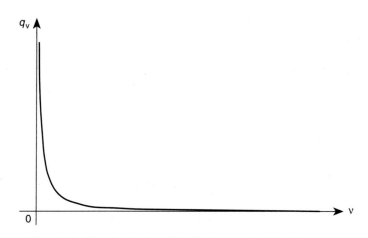

Figure 8.2 The vibrational partition function as a function of frequency.

Chemical background

In contrast to the translational and rotational partition functions, the vibrational partition function requires a consideration of both potential and kinetic energies, both being involved in this type of motion.

The vibrational frequency v corresponds to a particular normal mode of vibration, and the contributions to the partition function from each normal mode need to be added. For linear polyatomic molecules containing N atoms, there are $3N - 5$ normal modes, while this decreases to $3N - 6$ for nonlinear polyatomics.

Solution to worked example

A function such as this will only become discontinuous if we attempt to divide by zero. This can only happen if the equation

$$1 - \exp\left(\frac{-hv}{kT}\right) = 0$$

is satisfied. This would require

$$1 = \exp\left(\frac{-hv}{kT}\right)$$

Since we know that any number raised to the power zero is one, $e^0 = 1$, so that

$$\frac{hv}{kT} = 0$$

which can only be satisfied if the vibrational frequency v is zero. The graph of this function is shown in Figure 8.2.

8.4.2 Integration by substitution

In section 6.5.1, we met the technique of integration by parts, which was used for the integration of certain types of functions which could not be treated using the simple rules we had already learned. Other functions may be integrated by making an appropriate substitution.

Suppose we need to calculate the integral

$$\int_0^1 (x + 5)^4 \, dx$$

One way of doing this, of course, would be to expand the bracket using Pascal's Triangle as outlined in section 7.6.1, and then to integrate each term of the polynomial individually. However, a quicker method which is of more general use is as follows.

If we make the substitution

$$u = x + 5$$

we can differentiate this defining equation to give

$$\frac{du}{dx} = 1$$

since the constant 5 will differentiate to zero. We can then rearrange the expression for the derivative to give

$$dx = du$$

Since this is a definite integral, we can replace our limits in the variable x with the corresponding limits in the variable u. For the lower limit, when $x = 0$ we have

$$u = x + 5$$
$$= 0 + 5$$
$$= 5$$

For the upper limit, $x = 1$ and so

$$u = x + 5$$
$$= 1 + 5$$
$$= 6$$

We are now able to work through the original expression for the integral, replacing terms relating to x with those relating to u. This involves:

• replacing the lower limit $x = 0$ with $u = 5$
• replacing the upper limit $x = 1$ with $u = 6$
• replacing $(x + 5)^4$ with u^4
• replacing dx with du.

The result of these replacements is that the integral to be calculated becomes

$$\int_1^6 u^4 \, du$$

This is simple to calculate using the basic rule of raising the power by one and dividing by the new power. We then have

$$\int_1^6 u^4 \, du = \left[\frac{u^5}{5}\right]_5^6$$

Substituting the appropriate limits and using a calculator to evaluate the terms then gives

$$\int_1^6 u^4 \, du = \frac{6^5}{5} - \frac{5^5}{5}$$

$$= \frac{7776 - 3125}{5}$$

$$= \frac{4651}{5}$$

$$= 930.2$$

Worked example 8.6

The rotational partition function q_r of a molecule is given in terms of the rotational quantum number J as

$$q_r = \int_0^\infty (2J + 1) \exp\left(\frac{-J(J + 1)h^2}{8\pi^2 I kT}\right) dJ$$

where I is the moment of inertia, h is Planck's constant, k is Boltzmann's constant and T is the absolute temperature. The upper limit on the integral sign is infinity, denoted by ∞. Calculate this integral in terms of these quantities by using the substitution $x = J(J + 1)$.

Chemical background

There is no simple algebraic expression for the rotational energy levels of an asymmetric top molecule, so the calculation of its rotational partition function is not straightforward.

The fact that the rotational partition function is calculated using an integral rather than a summation is due to the fact that the rotational levels are closely spaced. We are effectively treating the rotational levels as being continuous rather than discretely spaced, and under these conditions it is appropriate to use the integral.

The actual expression for q_r for diatomic molecules contains an additional quantity σ known as the symmetry number. This takes the value 1 for heteronuclear diatomics (such as HCl) and 2 for homonuclear diatomics (such as Cl_2)

to allow for the fact that in the latter a rotation of 180° produces an indistinguishable orientation.

Solution to worked example

The substitution suggested in the question is clearly related to the exponential term in the expression we want to integrate. As a first step it is useful to remove the brackets so we have

$$x = J(J + 1) = J^2 + J$$

This can now be differentiated term by term, using the rule that we multiply by the power and reduce the power by one. This gives

$$\frac{dx}{dJ} = 2J + 1$$

We can now separate variables and integrate

$$\int dJ = \int \frac{dx}{(2J + 1)}$$

It is straightforward to change the lower limit in J to one in x. When $J = 0$ we have

$$x = 0^2 + 0 = 0$$

Similarly, when $J = \infty$ it follows that $x = \infty$. It is not really correct to use infinity as a number in the above equation, but it should be apparent that the two limits are the same.

We can now substitute for each term in the original expression. It is actually easier to leave the initial $(2J + 1)$ bracket as it is, but the other changes required are:

- to replace $J(J + 1)$ by x in the exponential term
- to replace dJ by

$$\frac{dx}{2J + 1}$$

It so happens that, in this example, the limits on the integral sign remain unchanged. The integral to be calculated in terms of x is therefore

$$q_r = \int_0^\infty (2J + 1) \exp\left(\frac{-xh^2}{8\pi^2 IkT}\right) \frac{dx}{(2J + 1)}$$

The terms in $2J + 1$ cancel top and bottom and we are left with

$$q_r = \int_0^\infty \exp\left(\frac{-xh^2}{8\pi^2 IkT}\right) dx$$

This expression looks more complicated than it really is, and it is actually quite straightforward to integrate. The only variable in the exponential term is x, so we can group the other quantities to form a constant A defined by

$$A = \frac{h^2}{8\pi^2 IkT}$$

We can now write a much simpler expression for the integral to be calculated, which is

$$q_r = \int_0^\infty e^{-Ax}\, dx$$

We saw in section 6.5.2 that when we integrate the exponential function we obtain the same function divided by any constants. In this case the constant is $-A$, so we obtain

$$q_r = \int_0^\infty e^{-Ax}\, dx$$

$$= \left[\frac{e^{-Ax}}{-A} \right]_0^\infty$$

Instead of dividing by $-A$ within the brackets, this term can be brought outside to give

$$q_r = -\frac{1}{A} \left[e^{-Ax} \right]_0^\infty$$

Before we actually evaluate this expression at the specified limits, it is worth realizing that since

$$a^{-n} = \frac{1}{a^n}$$

our expression for q_r can be written as

$$q_r = -\frac{1}{A} \left[\frac{1}{e^{Ax}} \right]_0^\infty$$

If we now consider what happens at the upper limit of infinity, we see that we are taking the reciprocal of the exponential of a very large number. The exponential of a very large number is itself very large, so the reciprocal of this will be very small and we will take it as zero. The lower limit is zero, so Ax will be zero and we need the reciprocal of e^0. Since any number raised to the power zero is one, the reciprocal of e^0 is also one and this is our lower limit. Substituting into our expression for q_r now gives us

$$q_r = -\frac{1}{A} [0 - 1]$$

$$= -\frac{1}{A} [-1]$$

The two negative signs multiply to give a positive sign, so we obtain

$$q_r = \frac{1}{A}$$

Since we chose to define A as

$$\frac{h^2}{8\pi^2 IkT}$$

we finally have

$$q_r = \frac{1}{A} = \frac{8\pi^2 IkT}{h^2}$$

Exercises

1. Explain how you could plot a straight line graph from the following functions, and what information could be obtained, if c, d and z are constants and the other quantities are variables:

    (a) $a = \dfrac{bc}{d} + 2$

    (b) $y = \frac{2}{3}x^3 z^2$

2. Evaluate these quantities *without* using a calculator.

    (a) $\dfrac{7!}{3!}$

    (b) $\dfrac{6!}{(5!)^2}$

3. Write each of these expressions in factorial form.

    (a) $6 \times 5 \times 4$

    (b) $\dfrac{13 \times 12}{5 \times 4 \times 3 \times 2}$

4. Write each of these expressions as a multiple of a single factorial quantity.

    (a) $6! + 7!$

    (b) $2(10!) + 3(7!)$

5. Show that

$$\frac{d}{dx}(\ln kx) = \frac{1}{x}$$

regardless of the value of k.

6. Differentiate the functions:

(a) $x^2 \ln x$

(b) $x^3 e^{2x}$

with respect to x.

7. Which of these three functions are discontinuous?

$$f(x) = \frac{x^2}{x - 3} \qquad \text{for } x > 0$$

$$g(x) = \frac{x + 2}{x + 1} \qquad \text{for } x > -1$$

$$h(x) = \frac{3}{e^x - 1} \qquad \text{for } x \geq 0$$

8. Where are the discontinuities in these functions?

(a) $f(x) = \dfrac{x}{x^2 - 1}$

(b) $g(x) = \dfrac{e^x}{2 - \ln x}$

9. Determine these integrals by using an appropriate substitution:

(a) $\displaystyle\int_0^1 (2x + 1)^4 \, dx$

(b) $\displaystyle\int x \sqrt{x^2 + 2} \, dx$

10. Determine the following integrals by using an appropriate substitution:

(a) $\displaystyle\int e^x \sqrt{e^x + 2} \, dx$

(b) $\displaystyle\int_0^2 x(x^2 + 1)^3 \, dx$

Problems

1. The number of complexions Ω of a system is given by the expression

$$\Omega = \frac{N!}{\prod_i n_i!}$$

where N is the total number of particles and n_i is the number of particles in level i. Obtain an expression for Ω when

(a) all N particles are in different energy levels, as shown below.

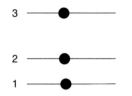

(b) all particles except one are in the lowest level, as shown below.

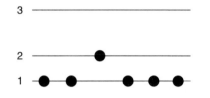

2. The translational partition function q_{trans} for a monatomic gas is given by the expression

$$q_{trans} = \frac{(2\pi mkT)^{3/2}V}{h^3}$$

where m is the mass of one molecule, k is Boltzmann's constant, T is the absolute temperature, V is the enclosing volume and h is Planck's constant. Obtain an expression for the pressure p of the gas by using the formula

$$p = NkT\left(\frac{\partial \ln q_{trans}}{\partial V}\right)_T$$

where N is the number of molecules present.

3. (a) Obtain the equation for Stirling's approximation by writing $\ln N!$ as a sum from $n = 1$ to $n = N$, replacing it by an integral which can be evaluated using the result of Question 10 in the Exercises in Chapter 6, and neglecting appropriate terms when N is large.
   (b) Rewrite the expression for Stirling's approximation in terms of an exponential rather than a logarithmic function.
   (c) What is the error introduced by using Stirling's approximation when $N = 10$?

4. The partition function q is defined by the equation

$$q = \sum_i \exp\left(\frac{-\varepsilon_i}{kT}\right)$$

where ε_i is the energy of level i, k is Boltzmann's constant and T is the absolute temperature. Show that if q is multiplied by a positive constant it gives q', which is the partition function for a system containing energy levels with the same spacing but a different zero.

5. When calculating the translational partition function for a monatomic gas, it is necessary to evaluate the integral

$$\int_0^\infty \exp\left(\frac{-n_x^2 h^2}{8ma^2 kT}\right) dn_x$$

where n_x is the quantum number for direction x, a is the corresponding dimension, m is the mass of the molecule, T is the absolute temperature, h is Planck's constant and k is Boltzmann's constant. This integral is of the general form

$$\int_0^\infty e^{-ax^2} dx$$

and its value can usually be obtained from appropriate tables. In this problem we will derive the result, however.

(a) If

$$I = \int_0^\infty e^{-ax^2} dx$$

show that

$$I^2 = \int_0^\infty \int_0^\infty e^{-a(x^2 + y^2)} dA$$

where $dA = dx\, dy$.

(b) This double integral can be transformed to polar coordinates r and θ using the equations

$$r^2 = x^2 + y^2$$

$$dA = r\, dr\, d\theta$$

Show that this gives

$$I^2 = \int_0^\infty r e^{-ar^2} dr \int_0^{\pi/2} d\theta$$

since r and θ are independent variables.

(c) Evaluate the integral

$$\int_0^\infty r e^{-ar^2} dr$$

by using the substitution $u = r^2$ and hence show that

$$I = \frac{1}{2}\left(\frac{\pi}{a}\right)^{1/2}$$

Fundamentals Appendix A

There are some mathematical processes without which you will find it very difficult to solve any problems, no matter how good your grasp of the other concepts involved. These are placed here so that you can find them easily if you need to refer to them while studying other parts of the book.

If we have a number represented by the algebraic symbol a, its negative will be written as $-a$. For example, the negative of 4 is -4. There are two rules to remember when combining numbers which may be positive or negative: two like symbols combine to give positive whereas two unlike symbols combine to give negative. This leads to the following sets of rules.

A.1.1 Addition

If you are adding a negative number, this is the same as subtracting the corresponding positive value. For example

$$2 + (-3) = 2 - 3 = -1$$

since the $+$ and $-$ signs are different and therefore give $-$. If our two numbers are represented algebraically as a and b then the rule can be written as

$$a + (-b) = a - b$$

A.1.2 Subtraction

If you are subtracting a negative number, this is equivalent to adding the corresponding positive number. For example

$$2 - (-4) = 2 + 4 = 6$$

since we have two like signs combining to give $+$. Using our symbols a and b we can write the general rule as

$$a - (-b) = a + b$$

A.1.3 Multiplication

When two positive numbers a and b are multiplied we simply obtain another positive number, as in

$$3 \times 4 = 12$$

We can represent this in symbols as

$$a \times b = ab$$

where it is usual to omit the multiplication sign $\times$.

If either number is negative (but not both) then the result will be negative, as in

$$-3 \times 4 = -12 \quad \text{and} \quad 3 \times -4 = -12$$

These are expressed in symbols as

$$-a \times b = -ab \quad \text{and} \quad a \times -b = -ab$$

If both numbers are negative, we have two like signs and so the result is positive, as shown by

$$-3 \times -4 = 12$$

The general rule is expressed as

$$-a \times -b = ab$$

A.1.4 Division

The rules for division are similar to those for multiplication. If 9 is divided by 3, we write

$$9 \div 3 = \frac{9}{3} = 3$$

Similarly, if a is divided by b we write this as

$$a \div b = \frac{a}{b}$$

If one of the numbers is negative we could have

$$-9 \div 3 = \frac{-9}{3} = -\frac{9}{3} = -3$$

and

$$9 \div (-3) = \frac{9}{-3} = -\frac{9}{3} = -3$$

since we are combining two different signs and therefore obtain a negative result. In general terms, this is written as

$$\frac{-a}{b} = \frac{a}{-b} = -\frac{a}{b}$$

On the other hand, if both numbers are negative, the two like signs give a positive result, as in

$$(-9) \div (-3) = \frac{-9}{-3} = \frac{9}{3} = 3$$

Using our symbols a and b this can be expressed as

$$(-a) \div (-b) = \frac{-a}{-b} = \frac{a}{b}$$

**A.2
Precedence in equations**

It is not always obvious in which order the different quantities in an equation need to be evaluated. However, this does not need to cause us any problems if we apply a simple set of rules:

- Evaluate quantities in brackets first.
- Raise numbers to any powers specified.
- Perform multiplications and divisions, working from left to right.
- Perform additions and subtractions, working from left to right.

Some examples should serve to make this idea clearer.

First consider the expression

$$2 \times 3 + 4 \times 5$$

The rules tell us to perform multiplication before division, so if we do this we have

$$2 \times 3 + 4 \times 5 = 6 + 20$$

which we can evaluate as 26. Now suppose that the expression given had been modified by the inclusion of brackets to become

$$2 \times (3 + 4) \times 5$$

We now have to evaluate the quantity in brackets first, and as this is equal to 7, we have

$$2 \times (3 + 4) \times 5 = 2 \times 7 \times 5$$

which has the value 70. Brackets can thus be used to override the standard order in which we would perform the operations.

An expression which involves raising to a power is

$$3 \times 2^5 - 1$$

The first operation is to apply the power, which gives us 32, we then multiply by 3 to give 96, and finally subtract 1 to give 95.

A.3
Rearranging equations   An equation can be thought of as a balance of two sides, each side consisting of terms which may be added, subtracted, multiplied or divided. For example, the equation

$$x + y = 25$$

contains two terms on the left-hand side and one on the right-hand side. Sometimes, we need to rewrite the equation in the form $y = \ldots$. If so, we need to rearrange the equation to make y the subject. To rearrange an equation while still keeping the balance of the equation, we can either add or subtract a term *from both sides*. Similarly we could multiply or divide both sides by the same number.

Suppose in the above equation that x has the value 10 and y the value 15. The equation then becomes

$$10 + 15 = 25$$

Using the rule given, we could subtract 10 from both sides to give

$$10 + 15 - 10 = 25 - 10$$

$$15 = 25 - 10$$

Notice that subtracting 10 from the left-hand side has the effect of removing the original 10, and leaving 15 on its own.

Similarly we could subtract 15 from both sides

$$10 + 15 - 15 = 25 - 15$$

$$10 = 25 - 15$$

Both of these statements are obviously correct, and the same types of rearrangements can be performed using the symbols. We might wish to rearrange the equation given above so that only y is on the left, so we subtract x from both sides to give

$$x + y - x = 25 - x$$

$$y = 25 - x$$

Certain terms in an equation may also contain a coefficient. For example in the expression

$$3y = 18$$

we would say that y has a coefficient of 3. To write this equation in the form $y = \ldots$, we need to divide both sides by 3. This can be seen by giving y the value of 6. We then have

$$3 \times 6 = 18$$

$$\frac{3 \times 6}{3} = \frac{18}{3}$$

$$6 = \frac{18}{3}$$

and similarly if we divide both sides by 6 we obtain

$$\frac{3 \times 6}{6} = \frac{18}{6}$$

$$3 = \frac{18}{6}$$

In a similar way, to isolate y in the equation given we need to divide both sides by the coefficient 3, so we obtain

$$3y = 18$$

$$\frac{3y}{3} = \frac{18}{3}$$

$$y = 6$$

For more complicated expressions involving the sum of terms with coefficients, we deal with the terms first and then the coefficients. For example, if

$$10x + 5y = 8$$

we can first isolate the term in y by subtracting $10x$ from both sides of the equation to give

$$10x + 5y - 10x = 8 - 10x$$

$$5y = 8 - 10x$$

We then divide both sides by the coefficient of y to obtain

$$\frac{5y}{5} = \frac{8 - 10x}{5}$$

$$y = \frac{8 - 10x}{5}$$

This can be tidied up slightly if required to give

$$y = \frac{8}{5} - 2x$$

A.4

Indices   If we have a quantity such as 2^4 we say that 2 is raised to the power 4 (or has an index of 4), and

$$2^4 = 2 \times 2 \times 2 \times 2$$

Here we have multiplied 2 together 4 times. In general, x^n is x raised to the power n, and consists of x multiplied together n times.

Some frequently encountered powers are given special names.

- x^2 is called 'x squared' and is equal to $x \times x$
- x^3 is called 'x cubed' and is equal to $x \times x \times x$.

There are a number of rules for combining numbers raised to powers.

A.4.1   Multiplication

When two numbers raised to powers are multiplied, we add the powers, so that

$$2^2 \times 2^3 = 2^{2+3} = 2^5$$

In general terms, this can be written as

$$x^a \times x^b = x^{a+b}$$

A.4.2   Division

When one number raised to a power is divided by the *same* number raised to another power, we take the difference between the powers, so that

$$\frac{2^5}{2^3} = 2^{5-3} = 2^2 = 4$$

In general terms,

$$\frac{x^a}{x^b} = x^{a-b}$$

A.4.3   Raising to a power

When a number which is itself raised to a power is raised to a power, we multiply the powers as in

$$(2^2)^3 = 2^{2 \times 3} = 2^6$$

This can be expressed algebraically as

$$(x^a)^b = x^{ab}$$

A.4.4   Roots

The square root of the number 16 is 4. This means that 4 multiplied by itself gives 16. It is also possible to calculate other roots, so that the root raised to the appropriate power gives back the appropriate number. For example, we can write that

$$\sqrt[3]{8} = 2$$

which states that the cube root of 8 is 2, and it follows that

$$2^3 = 8$$

Another way of expressing the cube root of 8 is as

$$\sqrt[3]{8} = 8^{1/3}$$

and in general, when the root of degree b is taken of a number x^a we simply divide the power a by power b, so that

$$\sqrt[b]{x^a} = x^{a/b}$$

A.4.5   Negative powers

A negative power simply denotes a reciprocal quantity ('one over' the number), so that

$$2^{-3} = \frac{1}{2^3}$$

which we express algebraically as

$$x^{-n} = \frac{1}{x^n}$$

**A.5
Fractions**

A fractional quantity is written as two numbers separated by a horizontal line. The fraction $\frac{4}{5}$ simply means 'divide 4 by 5'. This can also be written using the division sign as $4 \div 5$ and has the decimal value 0.8. So

$$4 \div 5 = \tfrac{4}{5} = 0.8$$

Algebraically,

$$\frac{a}{b}$$

simply means divide a by b, and may also be written as $a \div b$.

A.5.1   Identical fractions

If both the top and bottom of a fraction are multiplied by a constant its value remains unchanged:

$$\frac{1}{2} = \frac{2 \times 1}{2 \times 2} = \frac{2}{4}$$

or algebraically using symbols

$$\frac{a}{b} = \frac{c \times a}{c \times b} = \frac{ca}{cb}$$

Similarly, we can divide both top and bottom by a constant and the fraction remains unchanged:

$$\frac{3}{9} = \frac{\frac{3}{3}}{\frac{9}{3}} = \frac{1}{3}$$

In symbols, this is expressed as

$$\frac{ad}{bd} = \frac{\frac{ad}{d}}{\frac{bd}{d}} = \frac{a}{b}$$

A.5.2   Addition and subtraction

If we wish to add or subtract two fractions, we need to rewrite them with what is called a common denominator. This means that the number on the bottom of each fraction needs to be the same. For example, to calculate the sum

$$\frac{2}{3} + \frac{1}{4}$$

We cannot add the fractions until they both have the same denominator. Now, 3 and 4 are both factors of 12 (and 24 and 36 and …). We would need to write

$$\frac{2}{3} = \frac{2 \times 4}{3 \times 4} = \frac{8}{12}$$

and

$$\frac{1}{4} = \frac{1 \times 3}{4 \times 3} = \frac{3}{12}$$

to change $\frac{2}{3}$ and $\frac{1}{4}$ into their equivalent fractions $\frac{8}{12}$ and $\frac{3}{12}$ respectively. We can then evaluate

$$\frac{2}{3} + \frac{1}{4} = \frac{8}{12} + \frac{3}{12} = \frac{11}{12}$$

The best way of expressing this in symbols is

$$\frac{a}{b} + \frac{c}{d} = \frac{ad}{bd} + \frac{bc}{bd} = \frac{ad + bc}{bd}$$

It may be possible to simplify the expression obtained further, by dividing the top and the bottom by the same number. Subtraction of fractions is performed in a similar way.

A.5.3   Multiplication

When two fractions are multiplied, we can simply 'multiply across' the two fractions. The justification for this is that when multiplying by a fraction, it involves both a multiplication and a division. For example, in

$$\frac{2}{3} \times \frac{1}{4}$$

two-thirds of $\frac{1}{4}$ is two times one-third of $\frac{1}{4}$, so we could write

$$\frac{2}{3} \times \frac{1}{4} = 2 \times \frac{1}{3} \times \frac{1}{4}$$

A third of $\frac{1}{4}$ is the same as dividing $\frac{1}{4}$ by 3.

$$\frac{2}{3} \times \frac{1}{4} = 2 \times \frac{1}{3 \times 4} = \frac{2 \times 1}{3 \times 4} = \frac{2}{12}$$

In this case we can divide both top and bottom by 2 to give finally

$$\frac{2}{3} \times \frac{1}{4} = \frac{1}{6}$$

The algebraic expression for multiplication is

$$\frac{a}{b} \times \frac{c}{d} = \frac{ac}{bd}$$

A.5.4   Division

When dividing by a fraction it is the same as multiplying by its reciprocal (i.e. the fraction turned upside down). For example,

$$\frac{2}{3} \div \frac{1}{4} = \frac{2}{3} \times \frac{4}{1}$$

Dividing by $\frac{1}{4}$ is the same as asking 'how many quarters are there in …', and that is the same as multiplying by 4, the reciprocal of $\frac{1}{4}$. This can be written in general terms with symbols as

$$\frac{a}{b} \div \frac{c}{d} = \frac{a}{b} \times \frac{d}{c}$$

$$= \frac{ad}{bc}$$

A.6
Standard form   If a number is written as a value between 1 and 10 multiplied by a power of 10, it is said to be in standard form. This is useful for dealing with very small or very large numbers, as they can be represented in a concise way. Examples are

$$2.907 \times 10^3 = 2907$$

$$4.6 \times 10^{-4} = 0.00046$$

Notice that negative powers are used for numbers less than 1. A negative power 10^{-n} means that, in the normal form of the number the decimal point is n places to the left of the first significant digit. A positive power 10^n means that there are n places to the right of the first significant digit.

Units Appendix B

In modern scientific work we use the *Système International d'Unités*, which is more commonly known as the system of SI units. This is a coherent set of units which makes it straightforward to communicate physical quantities to other scientists.

B.1
Prefixes

Certain powers of 10 are given special names, which act as prefixes to the standard SI units. The ones you are likely to meet in chemistry are given in Table B.1.

Table B.1 Standard prefixes

Power of 10	Name	Symbol
10^6	mega	M
10^3	kilo	k
10^{-1}	deci	d
10^{-2}	centi	c
10^{-3}	milli	m
10^{-6}	micro	μ
10^{-9}	nano	n
10^{-12}	pico	p
10^{-15}	femto	f
10^{-18}	atto	a

So, for example, 1 nm = 1×10^{-9} m which is 1 nanometre. Notice that there is no space between the prefix and the symbol for the unit.

B.2
Equivalent units

When physical quantities are combined so are their units. Because of this, we need to know the relationships between units in order to be able to express the result of a calculation with its appropriate units. Some of the most common units you will meet can be expressed as follows:

hertz	$Hz = s^{-1}$
newton	$N = m\,kg\,s^{-2}$
pascal	$Pa = N\,m^{-2} = kg\,m^{-1}\,s^{-2}$
joule	$J = N\,m = m^2\,kg\,s^{-2}$
volt	$V = J\,C^{-1} = kg\,m^2\,s^{-3}\,A^{-1}$

Appendix C Physical constants

The values of some of the constants you will frequently meet when performing chemical calculations are given below.

Avogadro's constant	$L = 6.022 \times 10^{23} \text{ mol}^{-1}$
Boltzmann's constant	$k = 1.381 \times 10^{-23} \text{ J K}^{-1}$
Electronic charge	$e = 1.602 \times 10^{-19} \text{ C}$
Faraday constant	$F = 9.649 \times 10^{4} \text{ C mol}^{-1}$
Ideal gas constant	$R = 8.314 \text{ J K}^{-1} \text{ mol}^{-1}$
Planck's constant	$h = 6.626 \times 10^{-34} \text{ J s}$
Velocity of light in a vacuum	$c = 2.998 \times 10^{8} \text{ m s}^{-1}$

Answers to exercises

1. (a) 350   (b) 0.0012   (c) 0.54   (d) 13
2. (a) 13.85   (b) 0.25   (c) 99.54   (d) 0.01
3. (a) 78 500   (b) 0.00675   (c) 12.0   (d) 80 000
4. (a) 3.1   (b) 12.8   (c) 0.0   (d) 1.0
5. (a) 15.6   (b) 19.1   (c) 2.3   (d) 2.2   (e) 13.9
6. (a) 65.48   (b) 1160   (c) 1.24
7. (a) 108 cm^3 (b) 0.01   (c) 1%
8. $31.2 \pm 0.4 \text{ m s}^{-1}$
9. (a) 10.63   (b) 10.62   (c) 10.61
10. 0.17

1. (a) $x^3 y + x y^3$
   (b) $x^3 + x^2 y - x^2 y^2 - x y^3$
   (c) $x^2 - y^2$
2. (a) (i) 4   (ii) $f(-3) = 228$ (iii) $df/dx = 12x^3 + 6x^2 + 8x + 1$
   (b) (i) 6   (ii) $f(-3) = 5589$ (iii) $df/dx = 48x^5 + 5x^4$
   (c) (i) 3   (ii) $f(-3) = -215$ (iii) $df/dx = 24x^2$
3. (a) 380   (b) 1   (c) 1.099
4. (a) gradient = 6, intercept = 3
   (b) gradient = $\frac{5}{3}$, intercept = $\frac{4}{3}$
   (c) gradient = $\frac{1}{2}$, intercept = -4
5. proportionality constant = 3
6. $y = \dfrac{64}{x}$
7. (a) $f(1, -1) = -1$ (b) $f(0, 2) = 0$ (c) $f(-2, 1) = 16$
8. $\dfrac{\partial f}{\partial x} = 6x + 2y^2 + 12x^2 y^2, \dfrac{\partial f}{\partial y} = 4xy + 8x^3 y,$
   $df(x, y) = (6x + 2y^2 + 12x^2 y^2)\, dx + (4xy + 8x^3 y)\, dy$

9. $\dfrac{\partial^2 f}{\partial x\, \partial y} = \dfrac{\partial^2 f}{\partial y\, \partial x} = 4y + 24x^2 y$

10. (a) $x = 0.72$ or $x = -9.72$ (b) $x = -1.39$ or $x = -0.36$

Chapter 3   1. $p = 1.03$

2. $r = \dfrac{Kxy}{z^2}$

3. $x = 4.327$

4. (a) y against $\ln x$, gradient $= 6$, intercept $= 4$

(b) $\log y$ against $\sqrt{x}$, gradient $= 2$, intercept $= 0$

5. (a) $x = -3$ and $x = 3$ (b) $x = -\frac{3}{2}$

6. (a) $x = 0$ and $y = 0$ (b) $x = 0$ or $x = 1$

7. $x = 1, y = 0$

8. $x = \pm 2$

9. $x = \frac{1}{3}$, minimum

10. $x = \frac{1}{2}\sqrt[3]{\frac{1}{2}} = 0.397$

Chapter 4   1. (a) $2x^3 + \frac{9}{2}x^2 + 8x + C$

(b) $\frac{3}{4}x^4 + \frac{4}{3}x^3 + C$

2. (a) $\frac{2}{3}$ (b) 48

3. (a) $\frac{1}{2}x^2 + \ln x + C$ (b) 6.746

4. (a) $\log 12$ (b) $\ln 6$

5. $\dfrac{x}{(x + 2)(x + 3)} = \left(\dfrac{3}{x + 3}\right) - \left(\dfrac{2}{x + 2}\right)$

6. (a) $4x + \dfrac{3}{x}$

(b) $3 + \left(\dfrac{2x}{x^2 - 2}\right)$

7. (a) $\dfrac{x^2}{2} + \ln|x - 2| + C$ (b) 0.444

8. (a) 0.0996   (b) 29.56   (c) 3

9. (a) $\sqrt{\dfrac{x-5}{3}}$

(b) $\left(\dfrac{x+8}{2}\right)^2$

10. -0.0522

Chapter 5

1. 0.82, 0.59, 1.38

2. (a) 0.191   (b) 0.848   (c) 0.414

3. (a) $x = 20.6°$ (b) $x = 96.2°$

4. (a) $x = -0.280$ rad   (b) $x = -0.343$ rad

5. (a) 4.1   (b) 7.5

6. (a) $3\mathbf{i} + \mathbf{j} - 2\mathbf{k}$ (b) $-\mathbf{i} + 3\mathbf{j} - 4\mathbf{k}$

(c) -3 (d) $-\mathbf{i} - 7\mathbf{j} - 5\mathbf{k}$

7. 105.8°

8. (a) $\begin{pmatrix} 3 & 2 \\ 7 & 3 \end{pmatrix}$

(b) $\begin{pmatrix} 1 & 2 \\ -1 & -1 \end{pmatrix}$

(c) $\begin{pmatrix} 10 & 4 \\ 7 & 2 \end{pmatrix}$

9. $\begin{pmatrix} 7 \\ 10 \\ 6 \end{pmatrix}$

10. $\begin{pmatrix} 32 & 22 \\ 31 & 32 \end{pmatrix}$

Chapter 6

1. $xy = 10.1$

2. $f(x)$: minimum at $(1, 6)$; $g(x)$: minimum at $(1, 0)$

3. $z_1 z_2 = 6\left(\cos\dfrac{7\pi}{12} - i\sin\dfrac{7\pi}{12}\right)$

4. (a) 16, 19   (b) 37, 50

5. $f(n) = 3 \times 2^n$ for $n \geq 0$

6. $\operatorname{arc} f(x) = \sqrt{\dfrac{1}{a}\ln\left(\dfrac{x}{2}\right)}$

7. (a) $2x\ln x + x$ (b) $e^{2x}(2x + 1)$

8. (a) $3e^{3x+2}$

   (b) $\dfrac{(e^6 - 1)}{2} \approx 201.2$

9. $\dfrac{(e^2 + 1)}{4} \approx 2.097$

10. $x \ln x - x + C$

Chapter 7   1. (a) x inversely proportional to $\ln y$

(b) x^2 directly proportional to y

2. (a) y directly proportional to x, constant $= a/b$

(b) y inversely proportional to x^2, constant $= a^2 b$

3. (a) $\ln x$ (b) $\ln x^3 y$

4. (a) $\frac{1}{2}\ln(x + 1) - \frac{1}{2}\ln(x - 1)$

(b) $\ln 5 + 4 \ln x$

5. $f(2) - f(1) = 31$

6. $g(N + 1) - g(N) = 2N + 3$

7. 2

8. (a) $x^4 + 12x^3 + 54x^2 + 108x + 81$

(b) $32x^5 - 320x^4 + 1280x^3 - 2560x^2 + 2560x - 1024$

9. -108

10. 0.988 060

Chapter 8   1. (a) Plot a against b, gradient $= c/d$

(b) Plot y against x^3, gradient $= \frac{2}{3}z^2$

2. (a) 840   (b) $\frac{1}{20} = 0.05$

3. (a) $\dfrac{6!}{3!}$ (b) $\dfrac{13!}{5! \, 11!}$

4. (a) $8(6!)$ (b) $1443(7!)$

6. (a) $2x \ln x + x$ (b) $(3x^2 + 2x^3)\,e^{2x}$

7. $f(x)$ and $h(x)$

8. (a) $x = \pm 1$ (b) $x = e^2 \approx 7.389$

9. (a) 24.2   (b) $\frac{1}{3}(x^2 + 2)^{3/2} + C$

10. (a) $\frac{2}{3}(e^x + 2)^{3/2} + C$ (b) 78

Answers to problems

1. (a) (i) 1.54 Å   (ii) 1.54 Å
   (b) (i) 25.0°C   (ii) 25.01°C
   (c) (i) -433 kJ mol^{-1} (ii) -432.88 kJ mol^{-1}
   (d) (i) 16.0   (ii) 16.00
   (e) (i) 8.31 J K^{-1} mol^{-1} (ii) 8.31 J K^{-1} mol^{-1}

2. (a) 2.8 g cm^{-3} (b) 15.03 g
   (c) 159 kJ mol^{-1} (d) 7.9 kJ
   (e) 0.042 mol dm^{-3}

3. 0.12 J K^{-1} mol^{-1}

4. 5.64 ± 0.04 m^3

5. mean = median = 494 kJ mol^{-1}, mode = 493 kJ mol^{-1}, variance = 4 (kJ mol^{-1})^2, standard deviation = 2 kJ mol^{-1}

1. (a) nR/V (b) RT/V (c) RT/p

2. $\dfrac{\Delta H^{\ominus}}{RT^2}$

3. $\left(\dfrac{nR}{p}\right) dT - \left(\dfrac{nRT}{p^2}\right) dp$

4. $a = 0.138$ Pa m^6 mol^{-2}, $b = 3.18 \times 10^{-5}$ m^3 mol^{-1}

5. $V = \dfrac{RT \pm \sqrt{R^2 T^2 - 4ap}}{2p}$

1. (a) π against c, gradient = RT, intercept = 0
   (b) log k against $\sqrt{I}$, gradient = $1.02 z_A z_B$, intercept = log k_0
   (c) Λ against $\sqrt{c}$, gradient = $-(P + Q\Lambda_0)$, intercept = Λ_0

2. (a) $V_1 = \dfrac{1}{\rho}$ (b) $\Delta T_b = \dfrac{K_b m}{2}$

3. (a) Continuous, since $v > 1$

   (b) $\alpha = 1$, complete dissociation

4. 17.6 cm^3 mol^{-1}

5. 1005.65 cm^3

Chapter 4   1. 4.6×10^{-10} mol dm^{-3} s^{-1}

2. $\dfrac{d[NH_4Cl]}{dt} = -\dfrac{1}{3}\dfrac{d[Cl_2]}{dt} = -\dfrac{d[NCl_3]}{dt} = -\dfrac{1}{4}\dfrac{d[HCl]}{dt}$

3. (a) 4   (b) 2

4. First order, 6.61×10^{-6} s^{-1}

5. 81 kJ mol^{-1}

Chapter 5   1. 3.150 Å

2. 110.2°

3. $a\mathbf{i}, a\mathbf{j}, a\mathbf{k}, \frac{1}{2}a\mathbf{i} + \frac{1}{2}a\mathbf{j} + \frac{1}{2}a\mathbf{k}$

4. 364.3 Å^3

5. $\begin{pmatrix} x_2 \\ y_2 \end{pmatrix} = \begin{pmatrix} 1 & 0 \\ 0 & -1 \end{pmatrix}\begin{pmatrix} x_1 \\ y_1 \end{pmatrix}$

Chapter 6   1. Minimum at $r = \sqrt[3]{\dfrac{3A}{2B}}$

2. arc $E(n) = \dfrac{2d}{h}\sqrt{mn}$

3. $v - v_0 = 2.52 \times 10^{14} T_{max}$ where $v_0 = 5.47 \times 10^{14}$ Hz

4. arc $\psi_{1s}(r) = -\left(\dfrac{a_0}{Z}\right)\ln\left[\dfrac{r}{2}\left(\dfrac{a_0}{Z}\right)^{3/2}\right]$

5. $\dfrac{a^3\hbar^2}{6m}$

Chapter 7   1. $I = I_0\, e^{-bl}$

2. $F(J + 1) - F(J) = 2B(J + 1) - 4D(J + 1)^3$

3. $G(1) - G(0) = \omega_e - 2\omega_e x_e$

4. $3.813 - 0.0175(2m + 1) = 3.7955 - 0.035\, m$

5. The peak due to $(CH_3)_2$ will be split into two in the ratio $1:1$.
   The peak due to CHI will be split into seven in the ratio
   $1:6:15:20:15:6:1$

1. (a) $\Omega = \dfrac{N!}{1!1!1!...} = N!$

   (b) $\Omega = \dfrac{N!}{(N-1)!1!} = N$

2. $p = \dfrac{NkT}{V}$

3. (b) $N^N e^{-N}$ (c) 14%

4. $q' = \displaystyle\sum_i \exp\left(\dfrac{-(\varepsilon_i + \alpha)}{kT}\right)$

Chemical index

Mathematical index

$$\ln |xy| = \ln |x| + \ln |y|$$

$$\ln \left| \frac{x}{y} \right| = \ln |x| - \ln |y|$$

$$\ln |x^y| = y \ln |x|$$

$$PV = NRT$$

$$T = \frac{NR}{PV}$$

$$y = mx + c$$